AF327635

Modern Methods of
Particle Size Analysis

CHEMICAL ANALYSIS

A SERIES OF MONOGRAPHS ON ANALYTICAL CHEMISTRY AND ITS APPLICATIONS

Editors

P. J. ELVING, J. D. WINEFORDNER

Editor Emeritus: **I. M. KOLTHOFF**

Advisory Board

VOLUME 73

A WILEY-INTERSCIENCE PUBLICATION

JOHN WILEY & SONS

New York / Chichester / Brisbane / Toronto / Singapore

Modern Methods of Particle Size Analysis

Edited by
HOWARD G. BARTH

Hercules Incorporated
Wilmington, Delaware

A WILEY-INTERSCIENCE PUBLICATION

JOHN WILEY & SONS

New York / Chichester / Brisbane / Toronto / Singapore

Library of Congress Cataloging in Publication Data:

Main entry under title:

Modern methods of particle size analysis.

  (Chemical analysis, ISSN 0069-2883 ; v. 73)
  ''A Wiley-Interscience publication.''
  Includes index.
  1. Particle size determination.　I. Barth,
Howard G.　II. Series.

TA418.8.M63　1984　　620'.43'0287　　84-3630
ISBN 0-471-87571-6

Printed in the United States of America

10 9 8 7 6 5 4 3

CONTRIBUTORS

Lyle G. Bunville
Hercules Incorporated
Research Center
Wilmington, Delaware

Karin D. Caldwell
Department of Chemistry
University of Utah
Salt Lake City, Utah

Ben Chu
Chemistry Department
State University of New York
Stony Brook, New York

Art DiNapoli
Chemistry Department
State University of New York
Stony Brook, New York

Michael J. Groves
Pharmaceutics Department
College of Pharmacy
University of Illinois
Chicago, Illinois

Archie Hamielec
Department of Chemical Engineering
McMaster University
Hamilton, Ontario, Canada

David A. Hoagland
Department of Chemical Engineering
Princeton University
Princeton, New Jersey

Karen A. Larson
Department of Chemical Engineering
Princeton University
Princeton, New Jersey

Philip E. Plantz
Leeds and Northrup Instruments
MICROTRAC Products
St. Petersburg, Florida

Robert K. Prud'homme
Department of Chemical Engineering
Princeton University
Princeton, New Jersey

Bruce B. Weiner
Brookhaven Instruments Corporation
Ronkonkoma, New York

PREFACE

This book, written by experts in the field, brings together the latest developments in particle size analysis. The objective of this work is to present an overview of particle size methodologies and to focus on state-of-the-art advances.

As the size of particulate matter approaches the submicron region and goes into the molecular domain, it becomes increasingly difficult to determine particle dimensions. The instrumentation becomes more sophisticated, and the theoretical treatment of data becomes more difficult. For broad or complex particle size distributions, data interpretation and instrumentation is limited. These difficulties are enhanced for chemically heterogeneous samples. Although solutions to these problems are not easy, we hope that this book provides the reader with some guidance regarding available methodologies and data interpretation.

The first two chapters consist of general reviews of instrumentation for particle size analysis; the second chapter concentrates on techniques used to characterize submicron dispersions and emulsions. The theoretical background and operational principles of each technique are described, as well as advantages and limitations. An attempt was made to include most commercially available systems or to describe experimentally feasible approaches. These chapters should provide the reader with a broad overview of various techniques for particle size analysis and also serve as a guide in the selection of an appropriate system.

Chapters 3–6 deal with light scattering and diffraction methods for particle size analysis. Recent advances in these areas are described, including interpretation of results. Although not directly related to particle size analysis, Chapter 4 presents a transformation procedure using photon correlation spectroscopy that has been used to approximate the molecular weight distribution of polymers.

The application of chromatographic techniques to particle size analysis is described in the last three chapters. Chapter 7 reports on the use of field-flow fractionation techniques to obtain particle size distribution profiles of high resolution. Methods of converting detector response to particle concentration and techniques of correcting for peak dispersion are covered in Chapter 8. The application of hydrodynamic chromatography to characterize high molecular weight polymers is treated in Chapter 9.

I wish to express my sincere appreciation to all the contributing authors for their dedication and cooperation; their efforts made this volume possible. Also, I appreciate the suggestions and advice given by Drs. J. J. Kirkland and W. W. Yau of E. I. du Pont de Nemours & Co. and acknowledge the Federation of Analytical Chemistry and Spectroscopy Societies (FACSS) Governing Board for granting permission to include Chapters 1, 3–6, 8, and 9, which were presented in part at the 1981 8th Annual FACSS meeting in Philadelphia. Special thanks are due to Hercules Incorporated for giving me the opportunity and support to complete this project. Finally, I wish to express my indebtedness to my wife, Nedda, for her continual help in innumerable ways.

HOWARD G. BARTH

Wilmington, Delaware
June 1984

CONTENTS

CHAPTER

1

COMMERCIAL INSTRUMENTATION
FOR PARTICLE SIZE ANALYSIS

LYLE G. BUNVILLE

Hercules Incorporated
Research Center
Wilmington, Delaware

1. INTRODUCTION

A casual examination of current technical literature and trade publications
indicates that a wide variety of instrumentation is commercially available
for particle size analysis. This chapter reviews instrumentation that is
representative of the various physical phenomena utilized for particle size
analysis. For the most part, the instrumentation applies to dispersions of
particles in liquids and does not include techniques such as optical imaging, image analysis, and sieving.*

The ideal instrument for particle size analysis would undoubtedly provide a complete distribution function in terms of some characteristic dimension of the particle, cover a wide size range, and have minimum

* Manufacturers of commercially available instruments are listed in the Appendix.

ambiguity in the relationship of the instrument response function to the particle dimensions. An instrument requiring a minimum of additional information about other properties of the particle or the dispersion medium for interpretation of the data would also be highly desirable. A further, practical characteristic would be the requirement of minimum time for data generation and data analysis.

The wide variety of phenomena used in instrumentation for particle size analysis includes electrical properties (such as differential conductance), transport properties in the case of sedimentation, electrophoretic mobility and hydrodynamic chromatography, and optical properties represented by a variety of light scattering phenomena (such as time-averaged scattering, time or spatial fluctuations in scattered intensity, and turbidity). The modes of data generation by instrumentation exploiting these several phenomena range from the counting and sizing of individual particles as discrete events (e.g., Coulter Counter and HIAC PA-720) and the generation of data as a continuum of particle size distribution (sedimentation using the Micromeritics Sedigraphs or the Joyce Loebl Disc Centrifuge)
to techniques that generate only several averages of the distribution, as is the case with photon correlation spectroscopy (PCS). Instrumentation may be further classified as to whether calibration is required, using standard particle systems (Coulter Counter, HIAC PA-720), or is not required (sedimentation and PCS).

Instruments using the phenomena mentioned above generate complex data at a high rate, which requires that a means of data analysis be available to, or incorporated in, the instrument. In the latter case, the software may not be user modifiable, and the objectives of particle size analysis as interpreted by the instrument manufacturer may not be the same as those of the user. For example, most instruments with self-contained data-analysis systems present the results in terms of the equivalent spherical diameter (ESD), that is, the diameter of a sphere generating a detector response equivalent to that of an irregularly shaped particle. The assumption of a spherical shape is, however, not necessarily required for analysis of the data. The use of the detector response function (particle volume or particle cross-sectional area, for example) would be a more realistic representation of irregularly shaped particles.

2. ELECTRICAL PROPERTIES: RESISTIVE PULSE TECHNIQUE (COULTER PRINCIPLE)

In particle size analysis using the Coulter effect, if a nonconducting particle suspended in a conducting medium is placed within a small aperture,

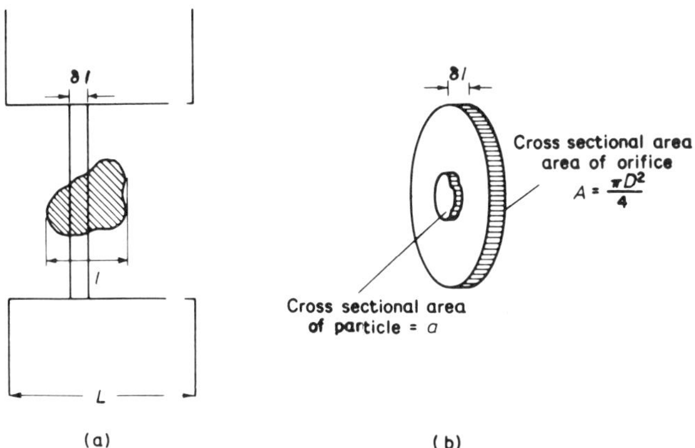

Figure 1. The passage of a particle through the orifice of a Coulter Counter; L and l are the lengths of the orifice and particle, respectively. An element of the particle, δl, in the orifice is shown in (b). From Allen (1). Used with permission of Chapman and Hall, Ltd., London.

an increase in the resistance across the orifice, relative to that of the medium alone is produced (Figure 1). The magnitude of this increase in resistance, ΔR, for a spherical particle of diameter d suspended in an aperture of diameter D, is

$$\Delta R = \frac{8\rho_f d}{3\pi D^4} \left[1 + \frac{4}{5}\left(\frac{d}{D}\right)^2 + \frac{24}{35}\left(\frac{d}{D}\right)^4 + \cdots \right] \tag{1}$$

where ρ_f is the resistivity of the conducting medium (1). With suitable external circuitry, this resistance pulse ΔR results in a voltage pulse $i\Delta R$ for a sphere of diameter d, where i is the current across the aperture. The resulting voltage pulses are counted and scaled using a multichannel analyzer. The schematic arrangement of the Coulter Counter is shown in Figure 2.

Instruments utilizing the resistive pulse technique require calibration using monodisperse particles with known diameter to assign a particle size to each of the thresholds. This takes into account, implicitly, the dimensions and electrical characteristics of the aperture and the conducting medium used. The dynamic range of a single aperture is $20\times$, and with a series of apertures the overall applicable size range is from approximately 0.5 to 1000 μm.

Typical of instrumentation available from Coulter Electronics, Inc. is the Model TAII, in which the distribution is resolved into 16 channels,

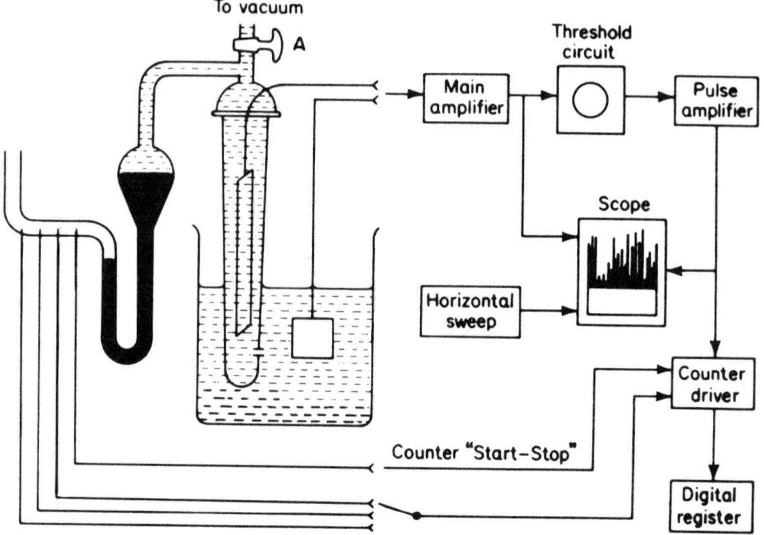

Figure 2.   Diagram of the Coulter Counter. Reproduced from Allen (1). By permission of Chapman and Hall Ltd., London.

the threshold of each channel having a common ratio of $2^{1/3} = 1.26$ (i.e., the ratio of the largest to the smallest diameter in each channel). The results of analysis are available as the particle volume fraction (normalized, integrated response), which is a function of the equivalent spherical diameter. By means of an accessory, the particle number distribution may be obtained.

Recent developments in instrumentation for particle size analysis using the resistive pulse technique include the incorporation of high resolution threshold circuitry. Up to 256 channels are available in the Particle Data, Inc., Elzone system.

A definite advantage of the resistive pulse technique is that no other property of the particle, such as refractive index or specific gravity, is required for interpretation of the data in terms of a particle size distribution. Limitations may arise, however, in the selection of carrier fluid with suitable conductivity and compatibility with the particular disperse phase under investigation.

An extension of the resistive pulse technique to particle sizes well below 1μm diameter was described by DeBlois and coworkers (2–4) and was used for measurements on viruses and other biological particles. This method, the Nanopar analyzer, is not yet commercialized, although it

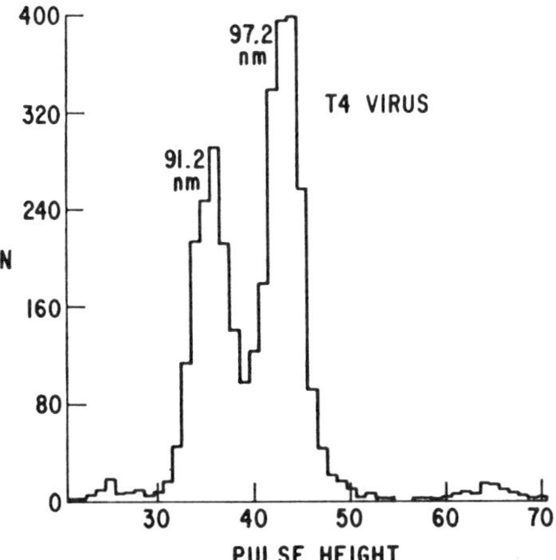

Figure 3. Pulse height histogram for T4 viruses with resistively equivalent diameters; N is the number of pulses with indicated height. DeBlois et al. (3). Used by permission of author and Academic Press, Inc., New York.

would be ideally suited to latices and emulsions, a size range where instrumentation for particle size distribution is limited (see Chapter 2). The essential difference between the Nanopar analyzer and other instrumentation for particle size analysis is that the sensor consists of a single pore in a Nuclepore (polycarbonate) membrane. The pore dimensions are tenths of a micron in diameter by 1–2 μm in length, and the membrane containing the isolated pore is mounted in a Plexiglas cell. (A schematic diagram of the Nanopar analyzer is shown in Chapter 2, Figure 23.) Virus particles of 0.06 μm have been analyzed and, based on the electrical characteristics of the technique, the Nanopar should be capable of sizing particles of 0.03 μm. Results for T4 viruses are shown in Figure 3. DeBlois et al. (4) compared the results obtained for insect viruses using the Nanopar with those obtained using PCS.

3. TRANSPORT PROPERTIES

3.1. Gravitational Sedimentation

The steady state velocity of a particle suspended in a viscous medium, settling in the gravitation field, is given by

$$M_e g = f\left(\frac{dx}{dt}\right) \tag{2}$$

where M_e is the particle effective mass, f is the frictional coefficient, dx/dt is the particle velocity, and g is the acceleration caused by gravity. Steady state velocities are customarily expressed as the sedimentation coefficient, S, the particle velocity per unit field. Thus, for Equation 2

$$S = \frac{1}{g}\frac{dx}{dt} = \frac{M_e}{f} \tag{3}$$

and the sedimentation coefficient is simply the ratio of the particle effective mass to the frictional coefficient. The extent to which sedimentation data can be interpreted in terms of particle size depends upon the particle shape and density and the density and viscosity of the medium. At the outset, the results from a sedimentation experiment can be given as a distribution of the sedimentation coefficient S. If the densities of the particle, ρ, and of the medium, ρ_0, are known, S can be expressed as

$$S = \frac{V(\rho - \rho_0)}{f} \tag{4}$$

where V is the particle volume. Finally, if the particles are spherical, with the frictional coefficient given by Stokes' law as $f = 3\pi\eta_0 d$ for a particle of diameter d suspended in a medium with viscosity η_0, the sedimentation coefficient becomes

$$S = \frac{(\rho - \rho_0)d^2}{18\eta_0} \tag{5}$$

from which the particle diameter d may be calculated. In the absence of this information, the sedimentation coefficient is generally interpreted as an equivalent spherical diameter, that is, the diameter of a sphere having the same sedimentation coefficient as the particle being analyzed.

For regularly shaped particles other than spheres, similar expressions can be obtained using frictional coefficients appropriate to the particle dimensions and geometry. For example, for cylindrical rods of length l and cross-sectional diameter d, the frictional coefficient for translational motion (5) can be used as

$$f_0 = \frac{3\pi\eta_0 l}{F_0(p)} \tag{6}$$

where $F_0(p)$ is a function of the axis ratio $p = l/d$, given by

$$F_0(p) = \ln 2p - 0.373 + 0.571p^{-1} \tag{7}$$

Using this in Equation 4

$$S = \frac{d^2(\rho - \rho_0)F_0(p)}{12\eta} \tag{8}$$

so that the sedimentation coefficient for cylinders is predominantly a function of the cylinder diameter.

Perhaps the most widely used instruments for particle size analysis based on gravitational sedimentation are the Micromeritics Sedigraph 5000-D and Sedigraph L. The two instruments differ in that the Sedigraph 5000-D uses the attenuation of soft x-rays to monitor particle concentration as a function of depth in the suspension, whereas the Sedigraph L uses white light as a concentration monitor.

A schematic diagram of the major components of the Sedigraph 5000-D is shown in Figure 4. At the start of an analysis, the sample cell is positioned relative to the stationary detector at a depth sufficient to include the largest particles expected in the sample. The cell is then moved downward through the x-ray beam at a programmed rate, so that the time and depth in the suspension define a given particle diameter. The attenuation of the x-ray beam monitors the concentration of disperse phase having this diameter. The results of analysis are obtained as a graph of the cumulative mass percent undersize as a function of equivalent spherical diameter. Results for the particle size distribution of NBS calibrated glass spheres are shown in Figure 5.

The overall size range that can be investigated by the Sedigraph is 0.1–100 μm, depending upon particle specific gravity. However, the applicibility to any specific particle system should be assessed in terms of the particle sedimentation coefficient, as discussed earlier. The dynamic range for a single analysis is 280×. Analysis times at the smaller size range of the Sedigraph can be long. For example, an analysis time of 190 minutes would be required for particles from 0.1 to 28 μm with specific gravity of 4, suspended in a medium with specific gravity of 1 and viscosity of 0.9 cp.

An attractive feature of the Sedigraph 5000-D is that the concentration monitor, x-ray absorption, gives a response that is independent of the size of particles in the disperse phase. For organic-based disperse phases, the x-ray absorption is too low for particle size analysis using the Sedigraph 5000-D. Even for disperse phases with suitable x-ray absorption coefficients, rather high disperse phase concentrations may be required to give suitable attenuation of the x-ray beam. Such high concentrations

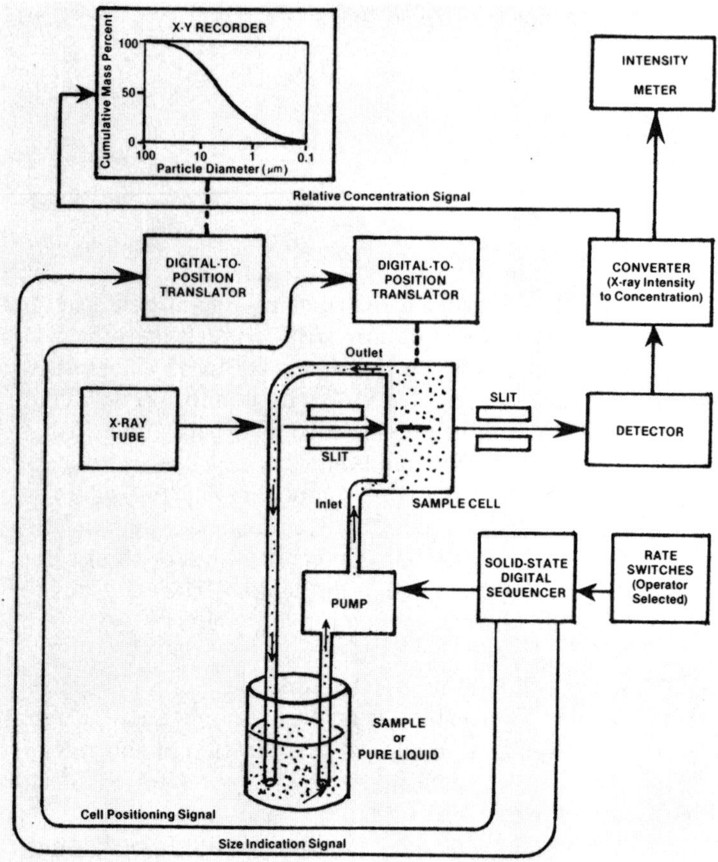

Figure 4. Operational schematic of the Sedigraph 5000-D. Courtesy of Micromeritics Instrument Corporation.

may lead to a low estimate of the particle size through concentration-dependent sedimentation or, for some types of disperse phases, may make the preparation of an adequately dispersed suspension difficult.

The Sedigraph L would be suitable for particle size analysis of disperse phases with low x-ray absorption, owing to the use of turbidity (white light) as a concentration monitor. The consequences of this are twofold: (1) a lower concentration of disperse phase may be used to carry out an analysis and (2) the detector response is no longer independent of the particle sizes in the disperse phase. The dependence of specific turbidity upon particle size and wavelength of incident light is illustrated in Figure

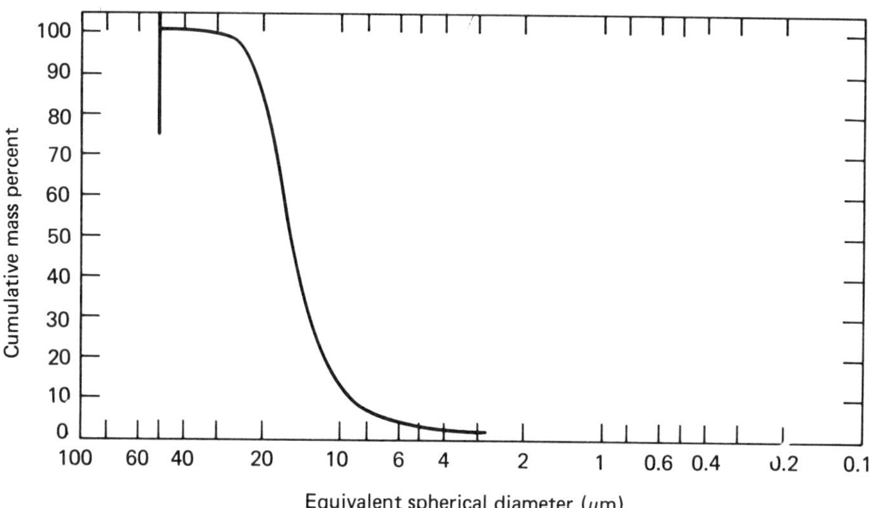

Figure 5. Particle size distribution of glass spheres, NBS SRM 1003, obtained using the Sedigraph 5000-D (circles are NBS results for the sample). Courtesy of Micromeritics Instrument Corporation.

6 (6), using a dimensionless size parameter $x = \pi n_0 (d/\lambda_0)$, where d is the particle diameter, λ_0 is the wavelength of the incident light, and n_0 is the refractive index of the medium. A value of x of 26 in water, at $\lambda_0 = 0.633$ μm corresponds to a particle diameter of approximately 4 μm. At larger particle diameters, the detector response varies approximately as the particle cross-sectional area.

Results of analyses using the Sedigraph L are shown in Figure 7 as the particle size distribution of milk chocolate solids. These results illustrate a feature common to both instruments, namely that even though particles with sizes smaller than 0.36 μm and below the range of analysis were present, the relative amount of these particles could be estimated in terms of area percentage. For example, the sweet chocolate contained 46% below 0.36 μm and the semisweet sample contained 16%.

A recently developed instrument for particle analysis, based on transport properties of the particles, is the Pen Kem System 3000 (7). The block diagram is shown in Figure 8, with the sample contained in a quartz tube, illuminated by laser radiation. The images of individual particles are focused on a rotating grating that intermitenly allows pulses of light to be transmitted to a signal detector at an angle of 90° to the incident beam. The major application to date of the System 3000 has been the determination of particle mobility. A potential is imposed across the col-

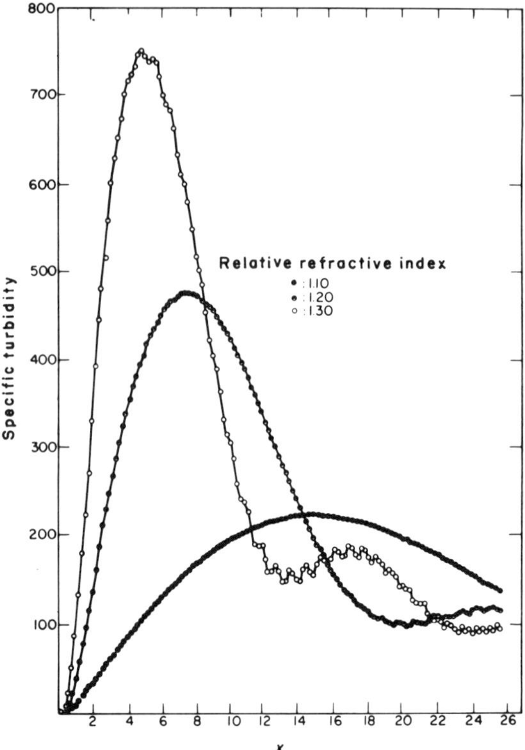

Figure 6. The specific turbudity as a function of size parameter x for a relative refractive index of 1.10, 1.20, and 1.30. From Kerker (6). By permission of the author and Academic Press, Inc., New York.

umn, resulting in a distribution of particle velocities that depends on the particle charge and frictional coefficient. As a result, a series of light pulses modulated by the rotating disk are seen by the detector. The duration of a particle pulse is determined by the particle velocity relative to the grating speed. The intensity of each pulse is determined by the 90° scattering properties of the particle. The output from the detector is processed by fast Fourier transform analysis to yield a histogram of the mobility distribution, with the weighting function being the scattered intensity of each particle (Figure 9).

The System 3000 can also be used to determine settling velocities in the absence of an applied field and, thereby, a response-weighted distribution of Stokes' law particle diameters. Figure 10 illustrates the velocity distribution for 1–3 μm glass spheres (settling velocities of the order of

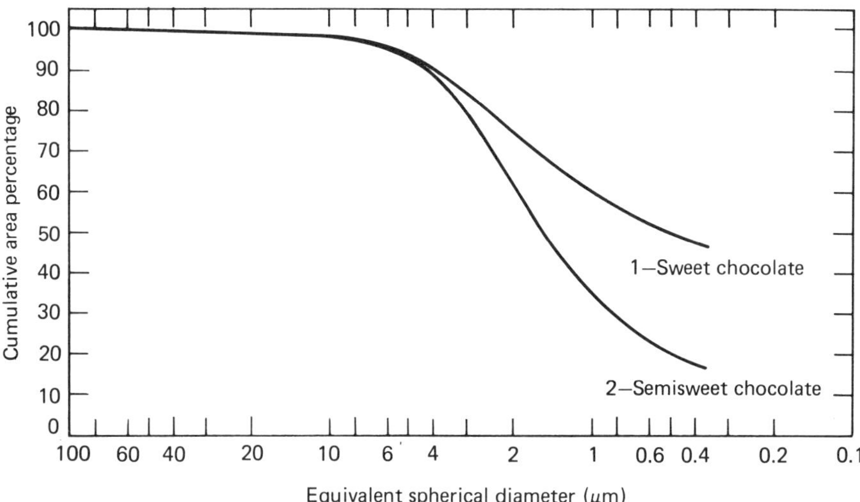

Figure 7. Particle size distribution of sweet and semisweet chocolate solids, obtained using the Sedigraph-L. Courtesy of Micromeritics Instrument Corporation.

4 μm/sec). The effects of Brownian motion can also be demonstrated. In this case, there is a random distribution of velocities and directions. Distributions for this effect are shown in Figure 11 for 0.5 and 0.1 μm particles.

3.2. Centrifugal Sedimentation

Instrumentation for particle size analysis using gravitational sedimentation is of limited flexibility. The only means of varying the particle velocity is by selecting a medium with different viscosity and specific gravity. By using the centrifugal field, a substantially lower size limit is achieved, along with a considerable decrease in the time required for analysis.

Equation 2, used earlier to describe the steady state velocity of a particle in the gravitational field, may be carried over to the centrifugal field by replacing the gravitational acceleration g by $\omega^2 r$, where ω is the angular velocity (radians/sec) and r is the radial distance of the particle from the axis of rotation. The result is

$$S = \frac{1}{\omega^2 r}\left(\frac{dr}{dt}\right) = \frac{M_e}{f} \tag{9}$$

The SA-CP2 series centrifuges offered by Shimadzu for particle size analysis represents a transition between gravitational and centrifugal sedi-

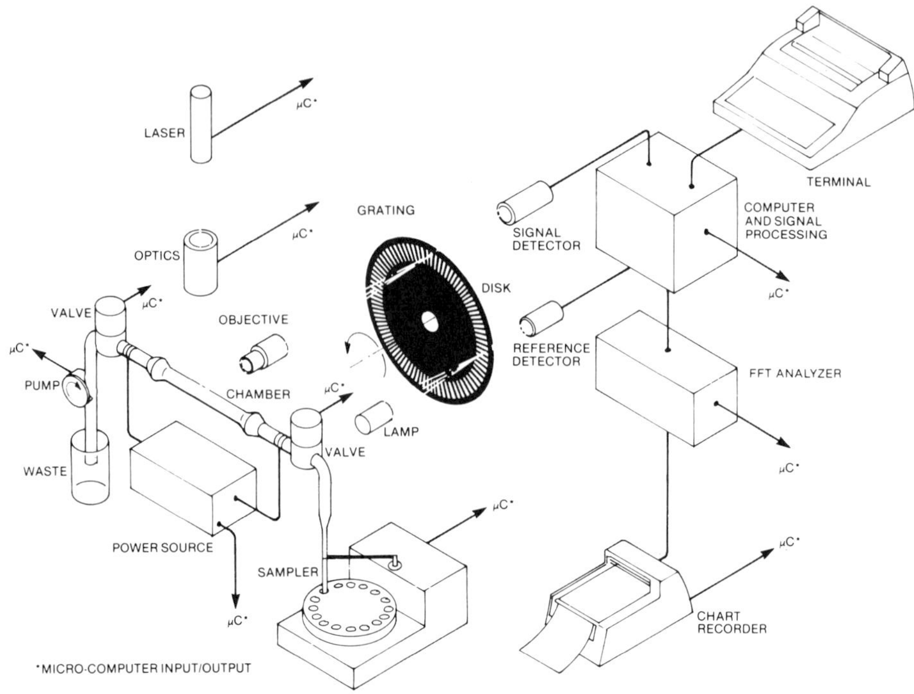

Figure 8. Block diagram of the Pen Kem System 3000. Courtesy of Pen Kem, Inc.

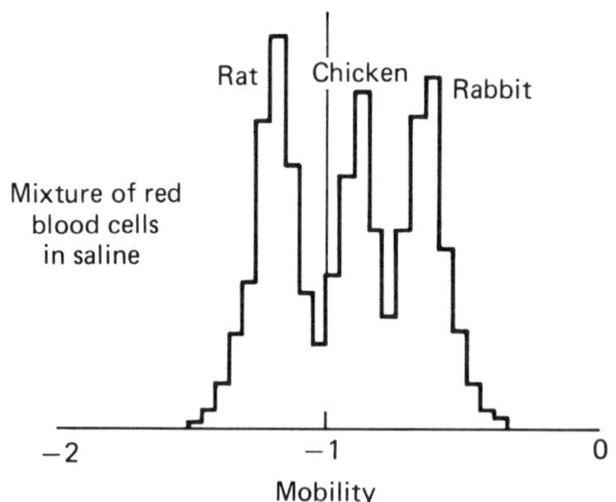

Figure 9. Mobility histogram of erythrocytes obtained using the Pen Kem System 3000. Courtesy of Pen Kem, Inc.

12

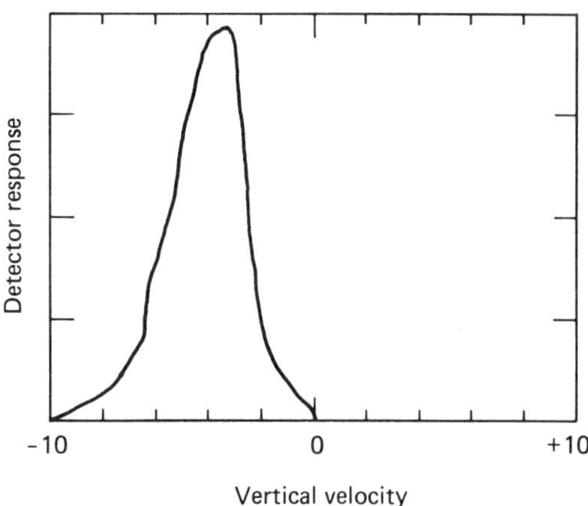

Figure 10. Histogram of settling velocities of 1–3 μm glass spheres, obtained using the Pen Kem System 3000. Courtesy of Pen Kem, Inc.

mentation. As shown in Figure 12, two cells are attached to the rotor; one is used as a counterbalance and one contains the sample dispersion. With the rotor stationary, gravitational sedimentation can be monitored (by an incandescent source with a filter) as a function of time; otherwise, centrifugal sedimentation may be monitored at speeds up to 1800 rpm. The SA-CP2-10 provides a data output of absorbance recorded as a func-

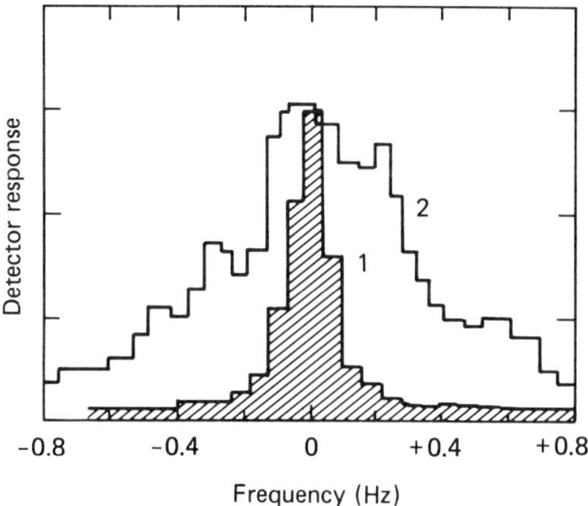

Figure 11. Spectra obtained for 0.5 μm diameter latex (curve 1) and 0.1 μm AgCl (curve 2) obtained using the Pen Kem System 3000 at zero field strength. Courtesy of Pen Kem, Inc.

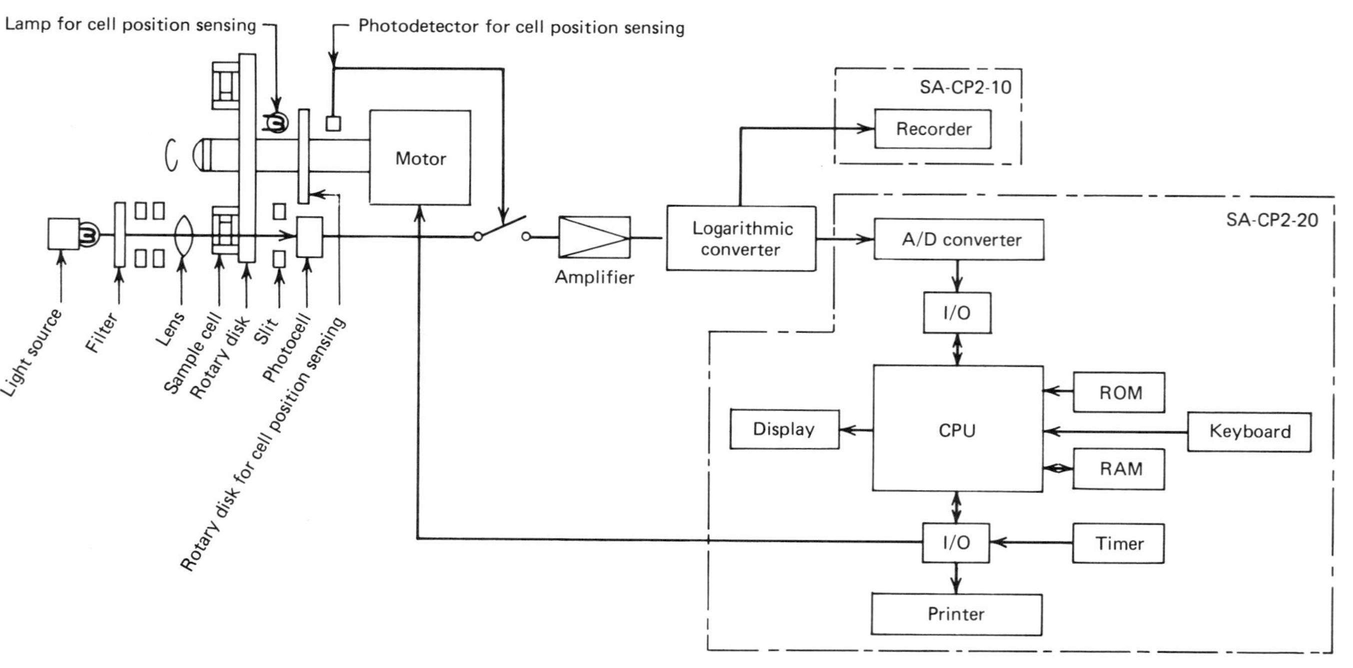

Figure 12. Block diagram of the Shimadzu Particle Size Analyzers SA-CP2-10 and SA-CP2-20. Courtesy of Shimadzu Scientific Instruments, Inc.

tion of time. The SA-CP2-20 has a microprocessor for digital conversion of the absorbance as a function of time to cumulative percent oversize as a function of particle diameter. Similar instrumentation for centrifugal particle size analysis is offered by Seishin, Union-Giken, and Horiba. The Model CAPA-500 available from Horiba Ltd. can be used for either gravitational sedimentation or centrifugal sedimentation at speeds up to 5000 rpm.

The instruments discussed so far for particle size analysis by sedimentation transport use the homogeneous suspension method: at the start of the analysis, the particle suspension is uniform throughout the cell. In gravitational sedimentation, at depths below the boundary, the particle concentration remains equal to the initial concentration, independent of both time and depth (except for the accumulation of particles at the base of the cell). Particle velocity remains constant throughout the analysis. In centrifugal sedimentation, the particle velocity increases linearly with increased radial distance from the axis of rotation. Because of this and the cylindrical geometry of the rotor, the particle concentration below the boundary, although remaining independent of position, decreases exponentially with time of sedimentation. This is referred to as the radial dilution effect and can be used to determine particle size (8). In centrifugal sedimentation analysis of homogeneous suspensions, this radial dilution effect complicates the mathematics required to obtain a particle size distribution from boundary analysis wherein particle concentration (or turbidity) is monitored as a function of time at a fixed radial depth in the suspension. Therefore, attention should be given to the nature of the mathematics and software used in data analysis.

An alternative method of particle size analysis by centrifugal sedimentation, capable of high resolution, is available with the Joyce–Loebl disk centrifuge photosedimentometer, which can achieve speeds up to 8000 rpm. The rotor, shown in Figure 13, is a hollow disk formed of transparent plastic. The scheme of analysis is the Joyce–Loebl buffered line start technique for zone transport analysis. Sedimentation fluid is loaded into the operating rotor, followed by a thin layer of a gradient fluid of lesser density. A buffer zone with a gradient of density is formed by both diffusion and the shearing action resulting from momentarily changing the speed of the rotor. Finally, a thin layer of the sample, dispersed in gradient fluid, is layered onto the buffer zone. The variously sized particles in the sample migrate through the sedimentation fluid in narrow bands or zones, with rates determined by their sedimentation coefficients. The time of transport from the injection radius to the fixed position of the detector is obtained from a strip chart recorder or microprocessor data acquisition system. The relative amount of each species is obtained

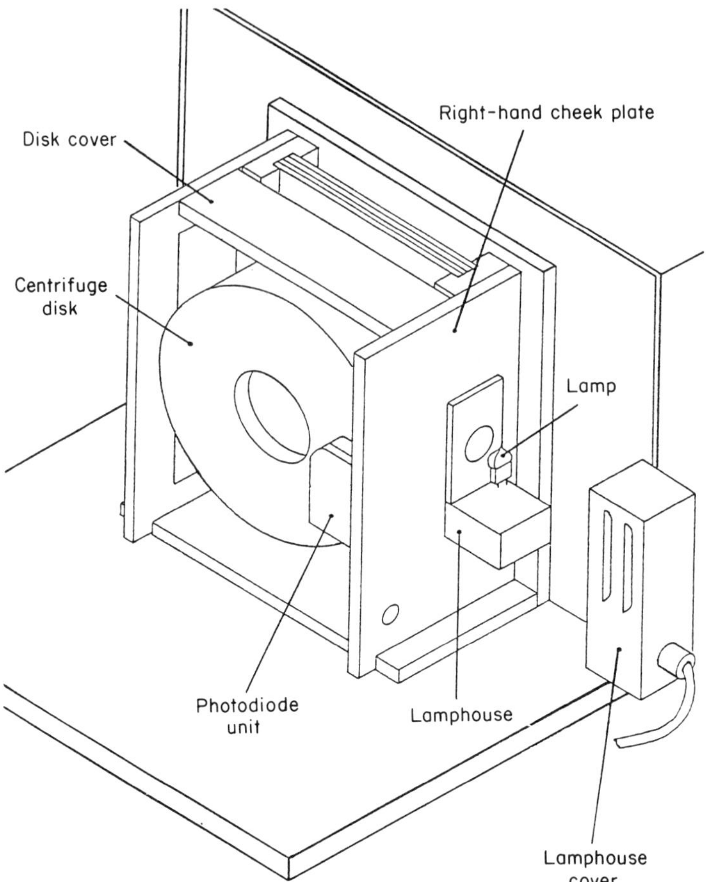

Figure 13. Sketch of the rotor used in the Joyce–Loebl Model DCF Mark III Disc Centrifuge Photosedimentometer. By permission of Joyce–Loebl.

from the detector response (white light turbidity) as a function of time. Typical results for a monodisperse polystyrene latex are shown in Figure 14 (9). After sonication of the sample, the expected unimodal distribution obtained was indicative of agglomeration in the original suspension.

A requirement for sedimentation transport is, of course, that the particle specific gravity be greater than that of the sedimentation fluid. However, to prevent convective transport or streaming of particles and suspension medium, it is essential that the injected sample suspension have a specific gravity less than that of the sedimentation fluid. Methanol–water mixtures work well for this purpose; not only is a positive density

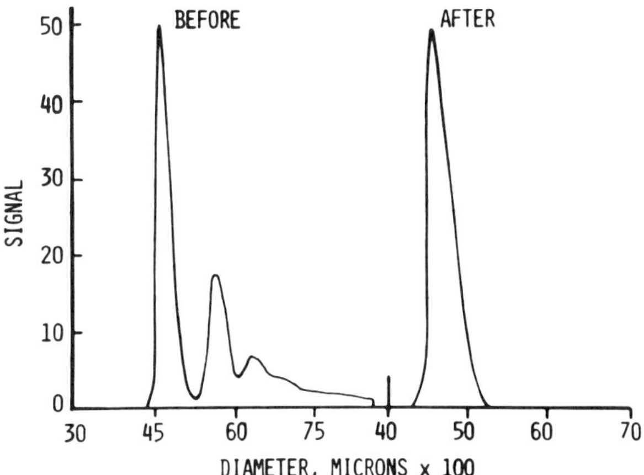

Figure 14. Particle size distribution of a monodisperse polystyrene latex obtained using the Joyce–Loebl disc centrifuge, before and after the use of ultrasound to disperse agglomerates. Detector response is in arbitrary units. From Bunville (9).

gradient produced in the gradient zone, but a negative viscosity gradient also results. This diminishes the tendency of the migrating particles to accumulate in the sedimentation fluid, which leads to convective instability.

A guide to the particle loading that can be injected without convective streaming of droplets of particles and suspending medium can be obtained as follows. Such droplets will have a velocity of

$$\frac{dr}{dt} = \frac{2\omega^2 r R^2(\rho - \rho_0)}{9\eta_0} \tag{10}$$

where R is the droplet radius and η_0 and ρ_0 are the viscosity and density of the particle-free medium at radial coordinate r (10). Theoretical considerations indicate that droplet streaming, rather than single particle transport, will occur when the number of particles per unit volume exceeds a limit proportional to

$$\frac{1}{a} \left[\frac{\omega^2 r}{D\eta_0} \cdot \frac{d\rho_0}{dr} \right]^{1/2} \tag{11}$$

where a is the particle radius and D is the diffusion coefficient of the suspended particle (11).

Overestimates of the width of the particle size distribution result when the zone transport method is used, not only from boundary spreading of small particles caused by diffusion, but also because of the finite width of the injected sample layer. Particles at the leading edge of the injected layer have a radial location slightly greater than those at the trailing edge. This, combined with the linear dependence of the particle velocity upon the radial coordinate in the rotor, results in a spectrum of arrival times, at the detector location, of particles having an identical diameter. Estimates of band broadening from this effect and from diffusion can be obtained using the relation

$$\sigma^2 = \left(\frac{\sigma_0 r}{r_0}\right)^2 + 2Dt\left(\frac{r}{r_0}\right) \tag{12}$$

where σ is the standard deviation of the band width at the detector radius r after sedimentation for time t, σ_0 is the standard deviation of the sample band initially injected at radius r_0, and D is the diffusion coefficient (12).

In summary, distributions of the sedimentation coefficient can be obtained from either gravitational or centrifugal sedimentation analysis. With additional information available as to the density and viscosity of the suspension medium, and the density and shape of the particle, the distribution of some characteristic dimension of the particle can be obtained.

3.3. Hydrodynamic Chromatography

Another instrument for particle size analysis based upon transport properties is the Flow Sizer 5600, which is being developed by Micromeritics and which utilizes hydrodynamic chromatography (HDC). A differential refractometer will be offered as a detector. The Flow Sizer 5600 will also incorporate a LSI-11 microcomputer for data analysis, including corrections for band broadening. The size range for packed bed HDC will be up to 1 μm, although capillary HDC is proposed as an optional accessory, with a size range from 1 to approximately 60 μm. If a suspension of particles is injected into a flowing carrier fluid (mobile phase) and then passed through a column packed with stationary, solid spheres, the larger particles elute from the column before the smaller particles (see Chapter 9). The reason is that larger particles in the flowing stream experience a larger mean velocity in the packed column, because their size excludes them from the slower carrier fluid velocity zones near the surface of the packing. This effect is illustrated in Figure 15, where the interstitial regions in the packed bed are visualized as short capillaries with a parabolic flow velocity profile (13). The contrast between the separation mechanism in HDC and SEC (size exclusion chromatography) is also illustrated; in

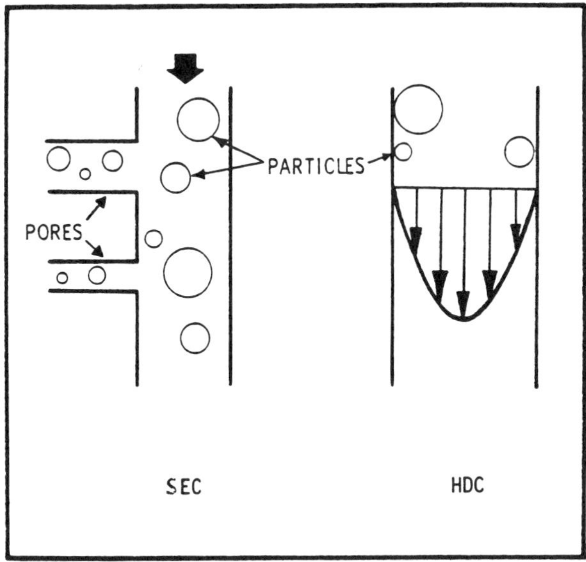

Figure 15.   Illustration of separation mechanisms in size exclusion chromatography (SEC) and hydrodynamic chromatography (HDC). After Nagy (13).

SEC, small particles (or polymers) are delayed by permeating porous regions in the packing. The use of HDC for particle size analysis was first demonstrated by Small (14). The equipment for implementing HDC is essentially the same as that employed in standard chromatographic systems. The separation achievable with HDC is illustrated in Figure 16 for 0.108 and 0.234 μm polystyrene latex spheres; a low molecular weight marker species, M, is included in the profile to define the elution volume.

Several potential problems with HDC are: (1) low sample recoveries (the packed bed may act as a filter for the larger particles in the distribution), (2) interactions between the particles and the packing, which may lead to adsorption of some species, (3) excessive sample dispersion within the column, and (4) the relatively low resolution of this technique. Even though appropriate instrumentation may be available for HDC, considerable effort might be required to develop a suitable mobile phase and packing for a given sample. Detector systems that might be used with HDC would be differential refractive index or absorbance (either single or multiwavelength). The former system gives a response more closely related to particle concentration, but requires higher sample loading. Absorbance (turbidity) has all the complexities arising from the dependence of light scattering upon particle size, shape, absorbance, and composition

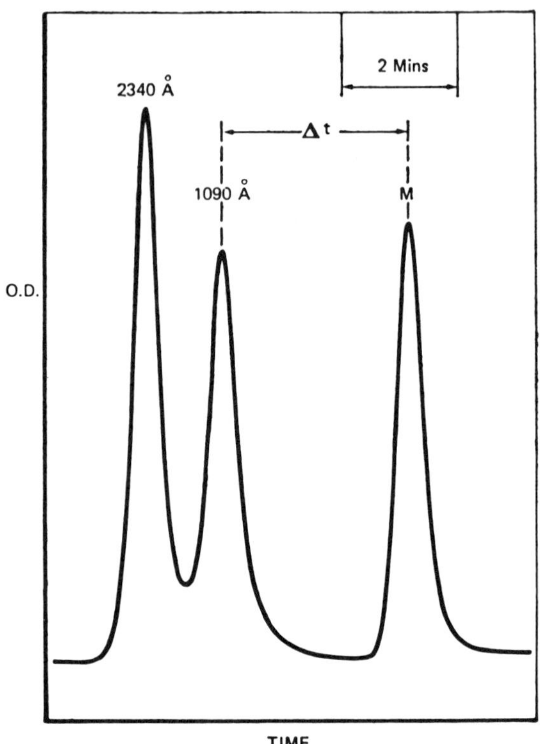

Figure 16. Typical HDC calibration chromatogram using 2340Å and 1090Å latex spheres. *M* is a marker solute used to define the elution volume. From Small (14). By permission of the author and Academic Press, Inc., New York.

(refractive index). See Chapter 8 for complete details regarding detector systems.

3.4.   Aerodynamic Transport

Aerodynamic transport is used for particle size analysis in the TSI Aerodynamic Particle Sizer APS 33. Airborne particles moving at a constant velocity are introduced through an orifice into a sheath air stream moving at a higher constant velocity. Upon entering this higher velocity air stream the particles accelerate, with the smaller particles accelerating more rapidly than the larger particles. The particles then pass individually through the detector zone comprised of a laser beam split into two parallel planes, each particle producing two scattering pulses detected by a photomultiplier in the forward direction of the inident beam. These two pulses thus define a time interval for the particle to move the fixed distance between the two parallel beams of the laser. The time intervals are stored in a multichannel accumulator for subsequent microcomputer data analysis. The size range of the APS 33 is 0.5–15 μm. A variety of output formats are available, including a 50-channel histogram of the particle number as a function of arodynamic diameter. The APS 33 requires calibration with particles of known diameter. The APS 33 would find strongest application in analysis of airborne particulates. However, aerosol generators are available for producing droplets containing single particles and evaporating the solvent for subsequent analysis by the APS 33. In fact, monodisperse polystyrene latices are used for calibration in the size range 0.5–2 μm (15).

4. NONIMAGING OPTICAL METHODS

4.1. Optical Blockage Technique

Perhaps the least complex optical phenomenon used for particle size analysis is that of light blockage, as incorporated in the HIAC series of instruments. In this method, the particles flow individually through an illuminated sensing zone, momentarily reducing the amount of light reaching a photodiode. The relation between the voltage output E of the photodiode and the particle cross section is given as

$$E = \frac{aE_0}{A} \tag{13}$$

where a is the particle cross section, A is the cross-sectional area of the sensor in the direction of the incident beam, and E_0 is the voltage from the photodiode in the absence of a particle. The configuration of the sensor is shown in Figure 17. The sensor is constructed so that turbulent flow is present. Thus, an irregular particle will give a variable signal as it tumbles through the sensing zone. The electronic circuitry of the HIAC instruments is such that the maximum output from the photodiode is detected, and the size characteristic of the particle is the maximum cross-sectional area. The pulse maxima are counted and sized using a multichannel analyzer. The optical blockage technique may be thought of as the optical analogue of the resistive pulse technique. A very distinct advantage of the optical blockage technique is that any carrier fluid compatible with the particle disperse phase may be used. For transparent

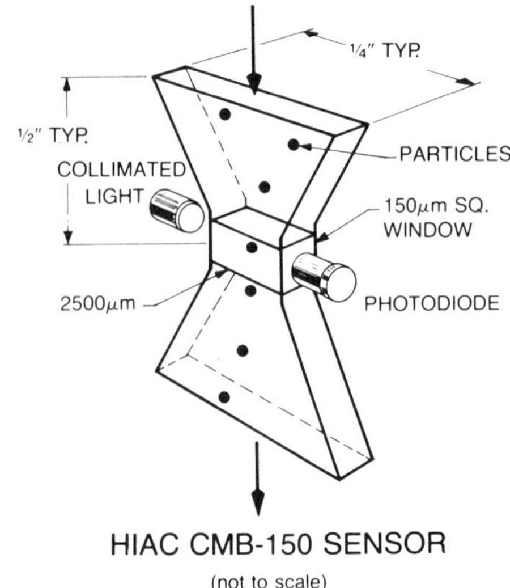

Figure 17.   Diagram of an HIAC optical pulse sensor. Courtesy of HIAC/Royco Instruments Division.

particles, the refractive index of the carrier fluid must differ from that of the particle.

With a series of sensors, the overall size range is from 1 to 9000 μm extending well into the sieve range. The dynamic range of a single sensor is ×45 for computation of particle size distribution, and up to ×60 for counting of oversize particles. The wide overall size range and wide dynamic range of a single sensor are distinct advantages in the analysis of broad distribution samples or samples with a trailing edge of contaminating, over-size particles.

A variety of HIAC instruments are available, ranging from 6- to 24-channel analysis. The instrument best suited for particle size distribution is the PA-720, with distribution into 23 channels and a common threshold ratio of $\sqrt{45}$ = 1.18 (i.e., the ratio of the largest to smallest diameter in each channel). Analysis results are available as the raw data (pulse count in each channel), population distribution, computed volume distribution (based on an equivalent sphere), and various averages of these distributions. The PA-720 will average the results of up to five replicate analyses (with up to 10^6 counts in each analysis) and compute the standard deviations for the normalized combined population distribution. The latter feature is useful in examining the sampling statistics of a given particle system.

Because of the direct nature of the response function (particle maximum cross-sectional area) of the HIAC sensor, the optical blockage technique would seem well-suited for the analysis of extremely irregular and ill-defined particles, as compared with the response function obtained in the diffraction techniques discussed in Section 4.2. With this system, the volume distribution is computed from the product of the particle count in each channel and the volume of a sphere having a diameter corresponding to the geometric mean diameter assigned to that channel. This is in contrast to the results obtained from the TAII Coulter Counter, which gives the total particle volume for each channel. A specific particle shape need not be assumed to compute apparent particle volume from the count data. Modification of the HIAC to provide an integrated response function (particle cross-sectional area) would require modification of the internal electronics of the PA-720. Nevertheless, for ill-defined irregular particles, a computed response distribution may be obtained from the population distribution (individual channel count) and calibrated mean cross-sectional area of each channel. Data analyzed in this manner does not require the assumption that the particles are spherical. This approach could be extended to gain some insight as to actual particle shape by comparing computed particle total volume, assuming spheres for example, and the actual volume of particles metered through the sensor.

4.2. Time-Averaged Light Scattering

Because of the complexity of light scattering and the number of ways it can be utilized for obtaining particle size information, a substantial number of instruments have been developed for this area. With these instruments, in addition to conventional light-scattering photometers, a mean size may be obtained from the angular independent scattering for sufficiently small particles ($d < 0.05$ μm). For somewhat larger particles (up to ~0.2 μm) two averages of the size distribution may be obtained from the angular dependence of the particle scattering factor and the scattering extrapolated to zero scattering angle. These techniques are discussed at length in Kerker's monograph (6).

Light-scattering phenomena can be classified into two areas, according to the mode of application to particle size analysis. The first category is time-averaged scattering, where either the scattered intensity or its spatial distribution is measured. The second is time-fluctuation scattering, which includes the analysis of the spectral distribution of the scattered radiation and PCS.

In the area of time-average scattering, two instruments that make use of both particle counting and sizing in terms of the intensity of scattered

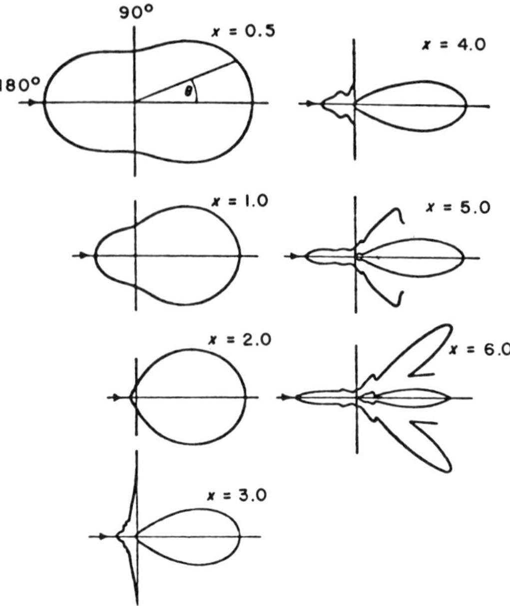

Figure 18.   Angular dependence of scattered intensity for relative refractive m = 1.55 and size parameter x from 0.5 to 6.0. The arrows indicate the direction of incident light. From Vouk (16).

light at low angles are the Spectrex Model ILI-1000 Laser Particle Counter and the HIAC-Royco Model 346. The complex nature of particle scattering is illustrated in Figure 18 as the scattered intensity as a function of angle for a size parameter up to x = 6 (corresponding to a particle diameter of approximately 1 μm at a wavelength of 0.633 μm in H_2O) (16).

The Spectrex ILI-1000 counts and sizes particles according to scattered intensity at an angle of ~15° to the incident beam. Analysis results are available as the number of particles (up to 999) in 1 ml of suspension, either above a present threshold between 1 and 100 μm, or as the number or weight histogram for the range 1–80 μm in 5 μm increments. The schematic of the optics bench is shown in Figure 19 and the detail of the scattering zone is illustrated in Figure 20. A volume of the suspension equivalent to 10 ml is scanned in 15 sec by the revolving laser beam, which sweeps out a cone-shaped path through the suspension. The advantages of the ILI-1000 are its ruggedness, portability, and applicability to samples in sealed containers. The latter is a distinct advantage in elim-

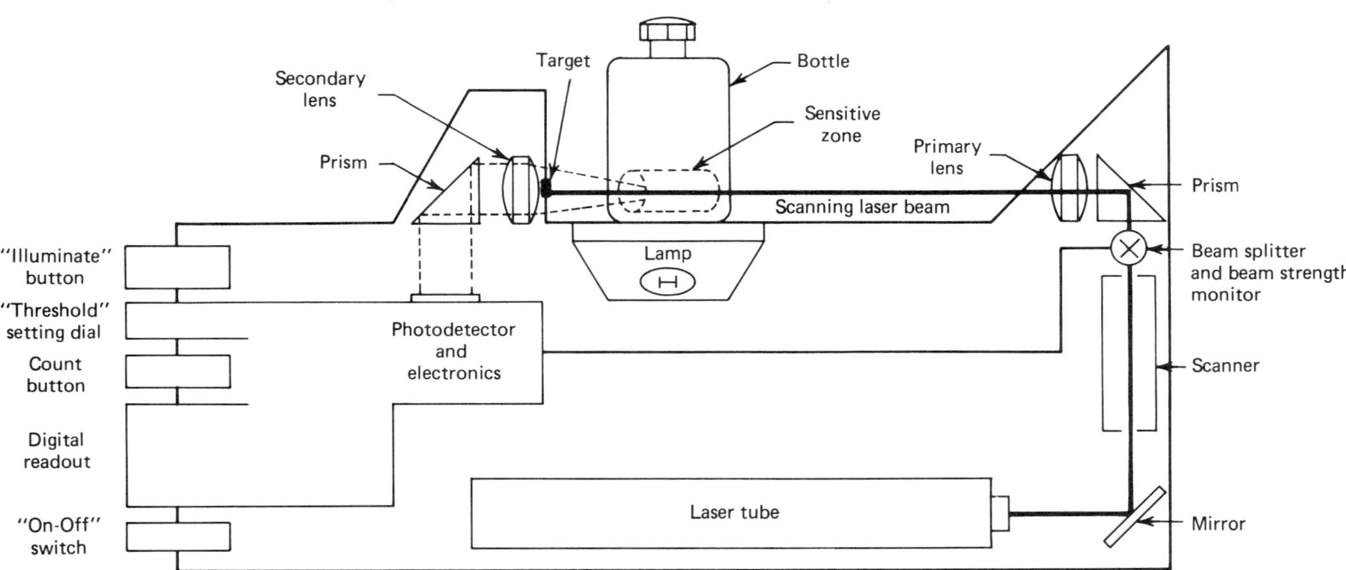

Figure 19. Schematic of the Spectrex ILI 1000 Laser Particle Counter. Courtesy of Spectrex Corporation.

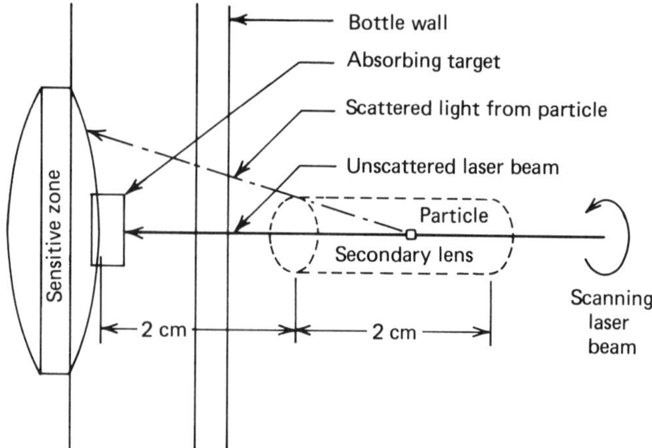

Figure 20. Schematic of the scattering zone of the Spectrex ILI 1000 Laser Particle Counter. Courtesy of Spectrex Corporation.

inating inadvertant contamination in sample handling or in the analysis of samples requiring isolation from the environment.

The recently introduced HIAC/Royco Model 346 sensor has an overall configuration similar to that used for optical blockage, but utilizes laser illumination (helium–neon) and detector optics responding to low-angle forward scattering (see Figure 6, Chapter 2). The overall size range is 0.5–25 μm and the sensor is compatible with several of the HIAC multichannel analyzers for data analysis. The Model 346 can be used for either on-line or batch sampling.

Several instruments are available for particle size analysis based upon the spatial distribution of time-averaged light scattering—that is, the analysis of the Fraunhofer diffraction pattern for moderate to large particles (see Chapters 5 and 6). For spheres, the spatial distribution of diffracted intensity is given by the Airy equation, which is discussed in detail in Chapter 6. The significant feature of particle size analysis by diffraction, in contrast to scattering, is that the intensity of diffracted light is not dependent upon the refractive index of the particle, because diffraction involves rays external to the particle. A particle system with a distribution of diameters will then give a composite diffraction pattern that is a summation of terms for each diameter. The nature of this effect is shown in Figure 21 for both a monodisperse and polydisperse assembly of particles (17). The monodisperse sample (3.9 μm diameter) gives rise to a series of sharp, concentric diffraction rings. The featureless appearance of the

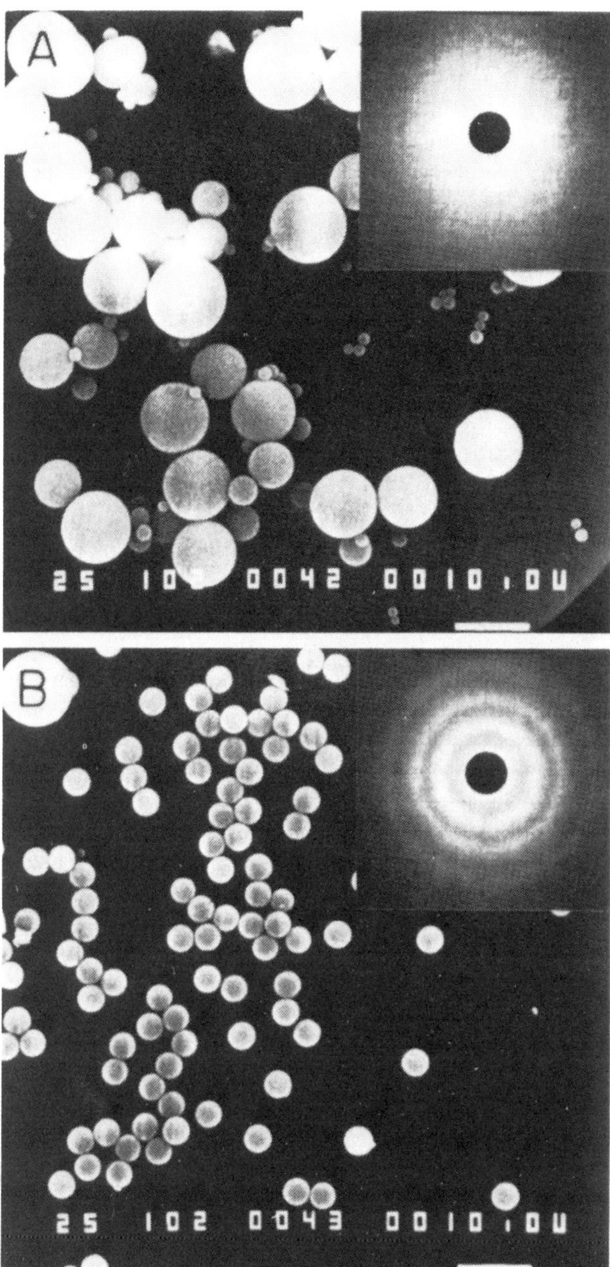

Figure 21. (*a*) Laser diffraction patterns and scanning electron micrographs for polydisperse spheres. (*b*) Monodisperse spheres with diameter 3.9 μm. From Almog et al. (17). Courtesy of the British Polymer Journal.

pattern from the polydisperse sample is the result of the superposition of many such rings, one set of rings with a different ring diameter for each particle diameter present in the sample.

Instruments utilizing Fraunhofer diffraction for particle size analysis differ mainly in the method used for analysis of the diffraction pattern. The Leeds and Northrup Microtrac instruments use specially designed rotating masks and apertures for analysis of the intensity distribution of the diffraction pattern (Chapter 6). The total flux diffracted by a sphere is proportional to the particle cross-sectional area, whereas the flux diffracted in a small limited central region of the pattern is proportional to the fourth power of the particle radius. By using a spatial optical filter that attenuates the diffracted flux as the inverse of the angle of diffraction, a detector response proportional to be the volume of the particle is obtained. Analysis results are available as a particle volume distribution in 13 channels. In addition to the standard 4-l stirred and pumped recirculating sample system, a small volume (250 ml) system and a powder sampler are available as options. Modifications of the Microtrac are available as in-line monitors. See Chapter 6 for further details.

The Cilas Model 75 Laser granulometer (Compagnie Industrielle des Lasers) uses a 15-element array detector in the optical focal plane for analysis of the intensity distribution of the diffraction pattern. The size range is 1–192 μm and analysis results are available as a cumulative or incremental weight percent or a light emitting diode matrix display of the particle size distribution.

The Malvern Systems, which use yet another data acquisition system for the intensity distribution, offer greater resolution for particle size analysis and greater flexibility in data analysis. For example, the Malvern Model 2200 Droplet Sizer (for sprays and aerosols) uses a solid–state detector comprised of 31 photoconductive concentric rings in the Fraunhofer plane (see Chapter 5). As a result, particle size distribution is available for an overall size range of 1–500 μm with three subranges, each subrange classified into 31 channels. With an accessory, the overall size range is expandable to 1800 μm in six subranges. The thresholds for the various ranges are user-modifiable. The results are available as incremental or cumulative weight percent on video display and hard copy. Further, any of four distribution functions may be selected for test modeling of the distribution results.

The Leeds and Northrup Small Particle Microtrac uses both Fraunhofer diffraction and light scattering for particle size analysis in the range 0.1–21 μm. Sizes and distributions from 0.34 to 21 μm are obtained by analysis of the intensity distribution of diffracted light, as in the standard Microtrac. Particle sizes from 0.1 to 0.34 μm are obtained from the 90°

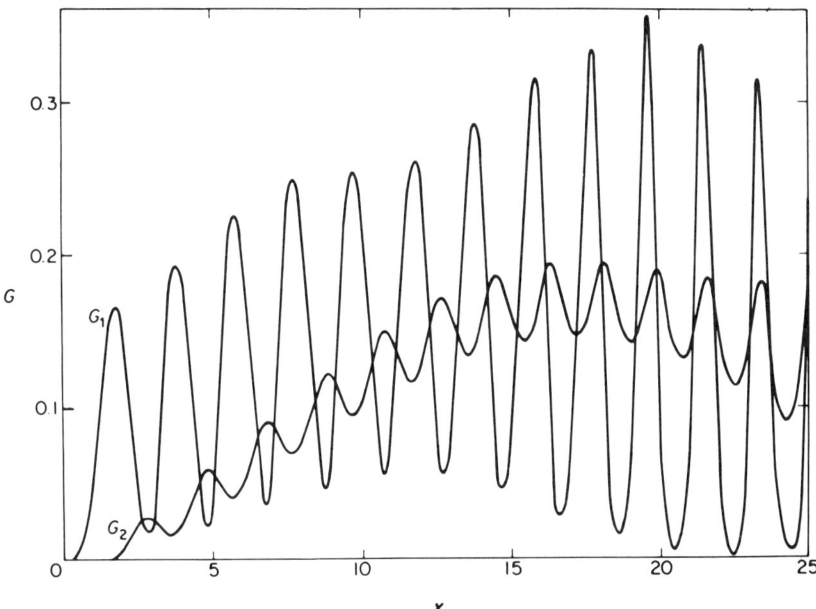

Figure 22. Angular gain for 90° scattering of incident vertically polarized (G_1) and horizontally polarized (G_2) light, as a function of particle size parameter x. From Kerker (6). By permission of the author and Academic Press, Inc., New York.

scattered intensity, using incident light of three wavelengths and both vertical and horizontal polarization. The optical arrangement of the Model 7991-3 is discussed in Chapter 6. The dependence of 90° scattered intensity upon particle size (expressed as the dimensionless size parameter x), and polarization of the incident light, is shown in Figure 22 for a particle with relative refractive index $m = 1.20$ ($m = n/n_0$, where n is the refractive index of the particle and n_0 is the refractive index of the medium). The scattered intensities are expressed as the angular gain functions $G_1 = 4i_1/x^2$ and $G_2 = 4i_2/x^2$, where i_1 and i_2 are the scattered intensities for unit intensity of vertically and horizontally polarized incident light, respectively. This phenomenon, expressed as the difference in scattered intensities per unit volume, $(i_1 - i_2)/x^3$, is the basis for size analysis for three channels in the range 0.1–0.34 μm in the Small Particle Microtrac (18).

As noted, the spatial distribution and intensity of diffracted radiation does not depend upon the particle refractive index. This, however, is not the case for scattered radiation, where the intensity of scattering varies

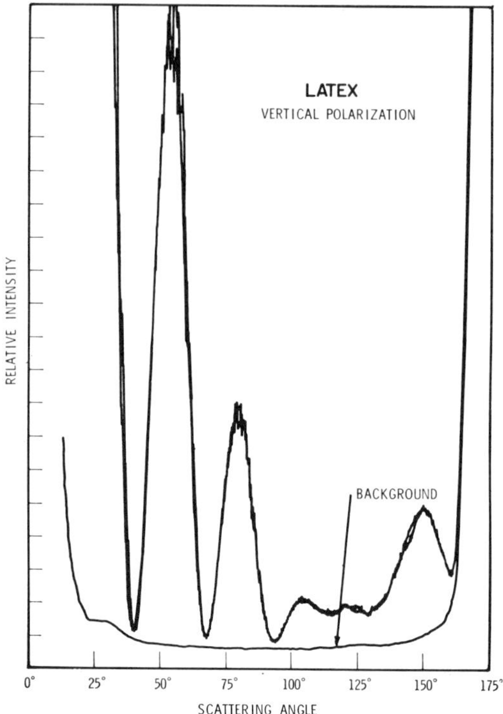

Figure 23.   Angular dependence of scattered intensity from a single polystyrene latex sphere (0.600 μm diameter), obtained using the Science Spectrum Differential II photometer. From Wyatt and Phillips (19). By permission of the authors and Academic Press, Inc. New York.

with refractive index. The Small Particle Microtrac uses a single value of the refractive index for computing particle sizes from polarized scattering intensities.

The complexities of the detector response function for light scattering and the capabilities of one photogoniometer for single particle analysis are illustrated in Figure 23. The scattering envelope shown for a single 0.6-μm polystyrene latex particle, nebulized and suspended in the incident beam in air, using an electrostatic field, was obtained using a Science Spectrum Differential II photometer (19).* This technique has been used for obtaining particle size distributions, thereby individually analyzing a large number of particles (20). Particle radii were obtained using Mie theory and the experimentally measured scattered intensities at 17 angles from 60° to 140° for each of 331 particles.

* This instrument is available from the Wyatt Technology Company.

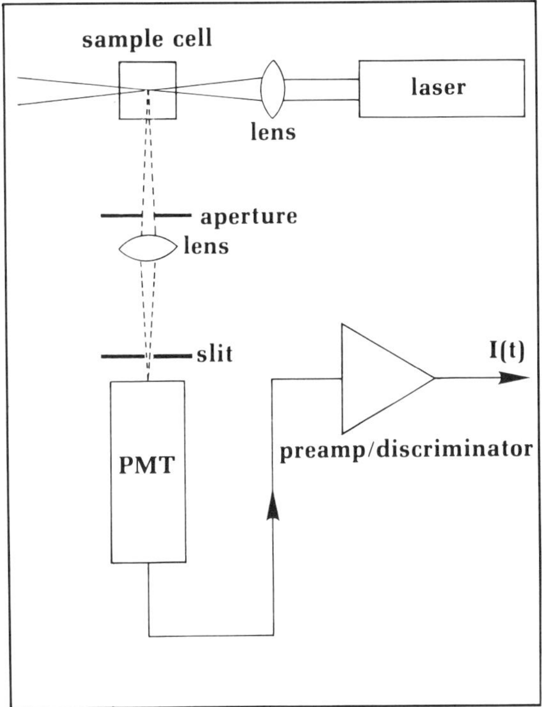

Figure 24. Schematic of optical bench for photon correlation spectroscopy. Courtesy of Nicomp Instruments.

4.3. Time-Dependent Light Scattering (Photon Correlation Spectroscopy)

Time-dependent light scattering, or PCS, is presently very much in the foreground of particle size analysis in the micron and submicron range. The hardware is quite similar to that used for conventional light scattering photometry, except that microscopic apertures are used to define a small volume of the suspension. A typical arrangement is shown in Figure 24. If the scattered intensity, or photocount, from this small scattering volume is measured repeatedly at sufficiently small time intervals, a signal similar to a noise signal is obtained. This signal and its origin are illustrated in Figure 25. At any instant in time, the scattered intensity is a function of the relative positions of the particles and the difference in optical path length to the detector. Minimum intensity is the result of destructive interference when this path length difference corresponds to $\lambda/2$. Maximum

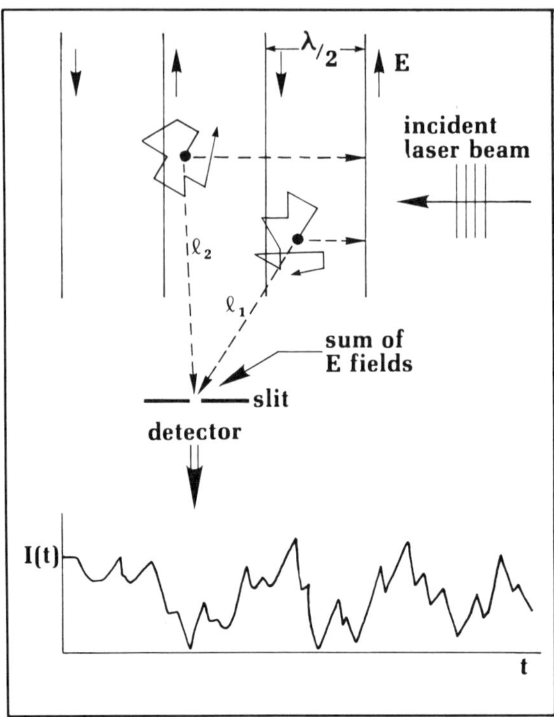

Figure 25.   Illustration of the origin of time-dependent scattered intensity and the corresponding fluctuating signal: E is the electric field of the incident beam; l_1 and l_2 are the distance of two particles from the detector; $I(t)$ is the fluctuation, with time, of the resulting signal. Courtesy of Nicomp Instruments.

intensity results from constructive interference when this path length difference corresponds to λ. The time dependence of this fluctuating intensity is governed by the random thermal or Brownian motion of the particles in a condensed phase, expressed as the diffusion coefficient. The data required for particle size analysis by PCS are values of the intensity (photocount), $I(t)$, at some arbitrary time, t, and at some later time, $I(t + \delta)$, where δ is the difference in time between the two counts. From these two photocounts, the autocorrelation function, $C(t)$, is computed as the product of the two intensities, $C(t) = [I(t){\cdot}I(t + \delta)]$. The correlation function is the running sum of such products for each of many delay times, typically from 32 to 124. The correlation function dependence upon time for a suspension of spheres monodisperse in size is given by

$$C(t) = A\exp(^-\Gamma t) + B \tag{14}$$

where B is the baseline (integrated intensity) and A is an instrument constant not required for data analysis. The baseline B corresponds to the square of the time-average scattered intensity: for sufficiently long delay time, all spatial distributions are equally probable and there is no correlation. The correlation function is most generally analyzed by the method of cumulants to obtain the decay coefficient Γ. This involves fitting $\ln C(t)$ to a power series in t as

$$\frac{1}{2}\ln[C(t) - B] = a_0 + a_1 t + a_2 t^2 + a_3 t^3 + \cdots \qquad (15)$$

and the mean decay time is $\Gamma = -a_1$. For a monodisperse suspension of spheres, the correlation function is an exponential function of time and coefficients a_2, and so on, will be negligible. The particle diffusion coefficient D, frictional coefficient f, and sphere diameter d are related as

$$\Gamma = 2K^2 D = 2\left[\left(\frac{4\pi n_0}{\lambda_0}\right)\sin\frac{\theta}{2}\right]^2 D \qquad (16)$$

$$D = \frac{kT}{f} \qquad (17)$$

$$f = 3\pi\eta d \qquad (18)$$

where k is Boltzman's constant and T is the temperature (°K).

Thus, from photon correlation analysis, a particle size may be obtained without requiring the use of any other property of the particle or of the concentration of the particle suspension. With regard to particle concentration, multiple scattering effects and effects from particle–particle interaction should be minimized. This can be done by measuring a series of concentrations and extrapolating the results to zero concentration. Further, the results of particle size analysis by PCS are not restricted to being expressed as a particle diameter. They could be expressed as a characteristic decay coefficient, diffusion coefficient, or frictional coefficient, according to the information available as to the conformation of the particle or the conditions of analysis. See Reference 21 and Chapter 3 for a more complete discussion of PCS for particle size analysis.

In the case of particle systems polydisperse with respect to size, the average size, though well-defined, incorporates the time-averaged or integrated scattered intensities as weighting functions. The result is complex, and particle size enters into the average in two ways: through the dependence of particle diffusion coefficient on size and through the size dependence of the particle-scattering functions. The correlation function is of the form

$$y(t) = \Sigma y_i(t) = \Sigma y_{io}\exp(-\Gamma_i t) \qquad (19)$$

where y_{io} is the weighting function (integrated intensity) for species i. By evaluating the first and second derivative of the correlation function at zero time, the average value and the standard deviation of the decay coefficient may be obtained, respectively, as

$$\bar{\Gamma}_y = \frac{\Sigma y_{io}\Gamma_i}{\Sigma y_{io}} \qquad (20)$$

$$\sigma_y^2 = \frac{\Sigma y_{io}\Gamma_i^2}{\Sigma y_{io}} - \left(\frac{\Sigma y_{io}\Gamma_i}{\Sigma y_{io}}\right)^2 = \bar{\Gamma_y^2} - \bar{\Gamma}_y^2 \qquad (21)$$

The subscript y indicates that these are intensity-weighted averages. A variety of scattering functions could be used to describe the dependence of the weighting functions on particle size and shape. For spheres, $y_{io} = N_i d_i^6 P_i(\theta)$, where N_i is the number of spheres with diameter d_i and particle scattering factor $P_i(\theta)$ at a scattering angle of θ (most commonly 90° for particle sizing instruments). Depending upon the size and size range of the spheres, a variety of expressions is available, beginning with the limiting case of $P(\theta) = 1$ (Rayleigh scattering) for sphere diameters not greater than $\sim \lambda/20$. For this limiting case, an average diameter for the distribution would be obtained as

$$\bar{d} = \frac{\Sigma N_i d_i^6}{\Sigma N_i d_i^5} \qquad (22)$$

If the particle size is such that intra-particle interference effects are not neglible, an average diameter is obtained as

$$\bar{d} = \frac{\Sigma N_i d_i^6 P_i}{\Sigma N_i d_i^5 P_i} \qquad (23)$$

For sizes not too much larger than Rayleigh scatterers, Guinier's Law (6) may be used as

$$P_i(\theta) = \exp(-K^2 r^2) = 1 - K^2 r^2 + \frac{K^4 r^4}{2!} - \cdots \qquad (24)$$

where r is the particle radius of gyration, $r = (3/20)^{1/2}\, d$.

For larger sizes, Rayleigh's expression (6) for the particle scattering factor can be used:

$$P_i(\theta) = \frac{9(\sin Ka - Ka \cos Ka)^2}{(Ka)^6} \tag{25}$$

where a is the particle radius, $a = d/2$.

In the completely general case, P_i would be obtained from the Mie theory; the complex dependence of the weighting functions upon particle size is indicated in Figure 22 as the 90° scattered intensity of incident, vertically polarized light. For this case, the average diameter obtained will depend somewhat on the refractive index of the particle, as well as on size, because of the dependence of the scattered intensity upon the dimensionless size parameter x.

The results obtained for asymmetric particles, such as rods, become even more complex. Rotational as well as translational diffusion effects are present even for a monodisperse system (all particles identical in size) and the preexponential factors, or weighting functions, depend in a complex way on particle size. The limiting form of the homodyne correlation function is

$$C_2(t) \sim \{R_0\exp(-\Gamma_0 t) + R_1\exp[-(\Gamma_0 + \Gamma_1)t]\}^2 \tag{26}$$

The preexponentials (weighting functions) R_0 and R_1 are the relative scattering intensities for translation and rotation, respectively; Γ_0 and Γ_1 are the decay coefficients for translational and rotational diffusion, respectively (22). A mean decay coefficient can then be obtained as

$$\bar{\Gamma} = \Gamma_0 + \frac{R_1\Gamma_1}{(R_0 + R_1)} \tag{27}$$

Further, for this monodisperse suspension of rods, the standard deviation of Γ is obtained as

$$\sigma^2 = \frac{R_0 R_1 \Gamma_1^2}{(R_0 + R_1)^2} \tag{28}$$

using Equation 21.

The relative magnitudes of the scattered intensities R_0 (translation) and R_1 (rotation) are shown in Figure 26 as a function of the size parameter $Kl = 4\pi\eta_0 l \sin\theta/2)/\lambda_0$ (23). This indicates that for small Kl (either small

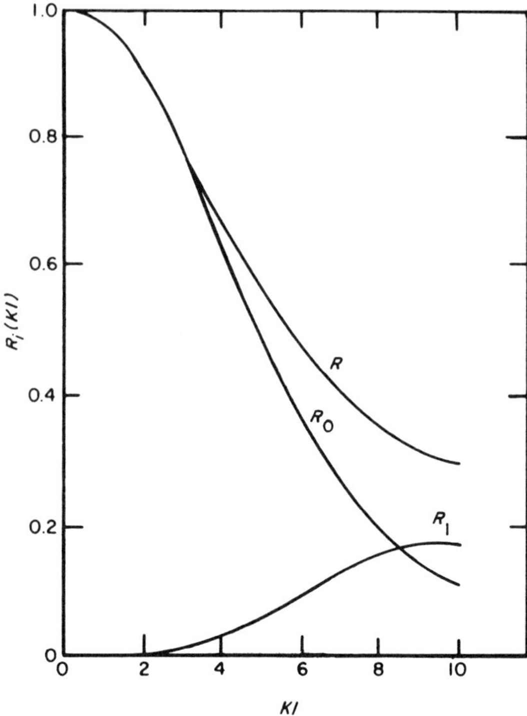

Figure 26. Relative integrated intensities of light scattered from isotropic rigid rods: R is the total relative integrated scattered intensity; R_0 is the intensity of the pure translational part; R_1 is the first nonzero term whose spectral width contains the rotational diffusion coefficient. From Pecora (23). Reproduced, with permission, from the *Annual Review of Biophysics and Bioengineering*, Volume 1, copyright 1972 by Annual Reviews Inc.

values of θ or l), the predominant contribution to the correlation function will be from translational diffusion, and that under these conditions, a particle length l could be obtained using the frictional coefficient as given in Equations 6 and 7. The decay coefficients used in Equations 26 and 27 are related to the cylinder dimensions as

$$\Gamma_0 = \frac{K^2 k T F_0(p)}{3\pi\eta l} \tag{29}$$

for translation, with $F_0(p)$ given by Equation 7. The corresponding quantity for rotation is

$$\Gamma_1 = \frac{18 k T F_1(p)}{\pi\eta l^3} \tag{30}$$

where $F_1(p) = \ln 2p - 1.45 + 7 (\ln^{-1} p - 0.27)^2$ (24); $p = l/d$ is the axis ratio for the cylinder. As a result, the mean "size" obtained for a suspension of rods can be a complex function of the particle dimensions.

The discussion above of the effect of particle shape (rods) on the results of particle size analysis by PCS illustrates three points: (1) for certain shapes, a suspension of dimensionally uniform particles may appear to be nonuniform; (2) for a suspension of rods, PCS instrumentation with multiangular capabilities may aid in characterization of the particles through the dependence of Γ_0 (translation) on angle and the angular independence of Γ_1 (rotation); (3) instrumentation with user-modifiable software would allow the results of analysis to be expressed in a manner more appropriate to the particle system under consideration.

The Nanosizer, marketed by Coulter Electronics, Inc., was an early entry in the area of instrumentation for particle size analysis by means of PCS. The applicable size range is 0.04–3 μm, with results obtained as a mean equivalent spherical diameter and a qualitative polydispersity index (PDI) ranging from 0 to 9, with increasing polydispersity. Although the PDI is qualitative, samples could at least be ranked according to distribution width. The Nanosizer obtains four values of the intensity correlation function, and from a linear least-squares fit of the log of the correlation function, the particle size is obtained. Since a linear fit is used whether or not the data are linear, the results are an underestimate of the particle mean diameter in the case of polydisperse systems. The PDI is estimated from the mean curvature of the log correlation function (25). The temperature of the analysis, refractive index, and viscosity of the suspension medium are entered into the microprocessor of the Nanosizer as a normalization factor, relative to a value of 100 for water at 20°C and computed as indicated by Equations 16–18. Analysis using the Nanosizer is carried out with a minimum of operator intervention. In the case of monodisperse suspensions, close agreement with the results of other methods was obtained for a large number of samples (26). The performance of the Nanosizer has also been investigated for polydisperse suspensions prepared as mixtures of characterized narrow distribution latices (27). The Nanosizer is no longer distributed by Coulter Electronics, Inc.

The Nicomp HN5-90/TC-100, Coulter N4, Brookhaven BI-90, Langley-Ford LSA-1, and Malvern instruments are representative of instrumentation which much more closely follows the requirements of PCS for particle size analysis. Typically, these instruments determine 32 (Coulter N4) or 64 (Nicomp) values of the correlation function and carry out a cumulants analysis for a first-order (linear), second-order (quadratic), and third-order (cubic) fit of the data, yielding for each fit the diffusion coefficient, the average particle diameter, the standard deviation of the size

Figure 27.   Brookhaven BI-90 particle sizer. Courtesy of Brookhaven Instruments Corporation.

distribution, and quantities related to the goodness of fit of the data analysis. These instruments vary in the extent to which operator interaction is required to determine analysis parameters, ranging from manual entry only, in the case of the Nicomp, to fully automatic, in the case of the Coulter N4 (except for viscosity, refractive index, and temperature) and the Brookhaven BI-90. The results of analysis can be obtained from both a video display and a printer. The instrument configuration of the Brookhaven Instruments BI-90 Particle Sizer is shown in Figure 27. The BI-90 incorporates an optional "dust-rejection" scheme of data acquisition and analysis to accommodate the inevitable presence of dust in routine analysis. The slow migration of dust particles into and out of the scattering volume results occasionally in abnormally high counting rates. The BI-90 "dust-rejection" mode of operation accumulates data in short batches, rejecting those with high counting rates and pooling the remaining normal batches.

Further differences in the various instruments appear in the area of data interpretation, that is, the presentation of a particle size distribution curve using the experimentally determined mean size and standard deviation and an assumed two-parameter distribution function. The Brookhaven BI-90 constructs such a curve using a log-normal distribution, and the Coulter N4 uses a Pearson Type V distribution. The resulting distribution is, of course, valid for the actual sample only if that sample follows the assumed distribution function.

Methods of data acquisition and analysis (i.e., correlator and computer technology) are presently in a state of flux, with improved hardware and software capabilities beginning to appear at the level of instrumentation referred to as particle-sizers, as differentiated from instrumentation for research in PCS. These advances should accommodate the long range of decay times present in polydisperse samples and the problem of deconvolution of the multiexponential correlation function. In the area of correlator technology are the multi-tau feature of the Langley-Ford 1096 correlator, which allows simultaneous development of correlation functions at three different delay times using the same data base, and the Malvern Loglin correlator, which allows the correlation function to be developed using either a linear or logarithmic time base. Although detailed particle size distributions have been available for some time, using Chu's histogram method of analysis of PCS data (see Chapter 4), only recently have detailed analysis methods for size distribution been incorporated in commercial particle sizers. Two examples are the loglin analysis in the Malvern 4600 system and the bimodal analysis in the Nicomp Model 200. The type of analysis available in the Nicomp 200 is illustrated in Figure 28 for a 3:1 mass ratio of 0.091 and 0.261 μm latex particles. The resolved experimental distribution is shown as the distribution of reciprocal diameter, and the peak sizes correspond to 0.092 and 0.279 μm with relative peak heights of 2.9 to 1.

The characteristics of commercial instruments for particle size analysis using PCS can be summarized as follows: they use a single scattering angle (90°); they most generally carry out a cumulants analysis for the mean and standard deviation of the size distribution; in some cases, they construct a size distribution curve using a two-parameter distribution function (Coulter N4 and Brookhaven BI-90); in one instance, carry out a bimodal data analysis (Nicomp 200). However, for research purposes, customized instrumentation for particle size analysis using PCS could be assembled using multiangular photometers (Malvern, Brookhaven Instruments), low angle photometry (Chromatix KMX-6), and any of a wide variety of correlator/computer configurations available from such companies as Nicomp, Langley-Ford, Malvern, and Brookhaven Instruments.

5. SUMMARY

It is apparent from this chapter that a wide variety of phenomena have been used by a broad range of commercially available instruments for

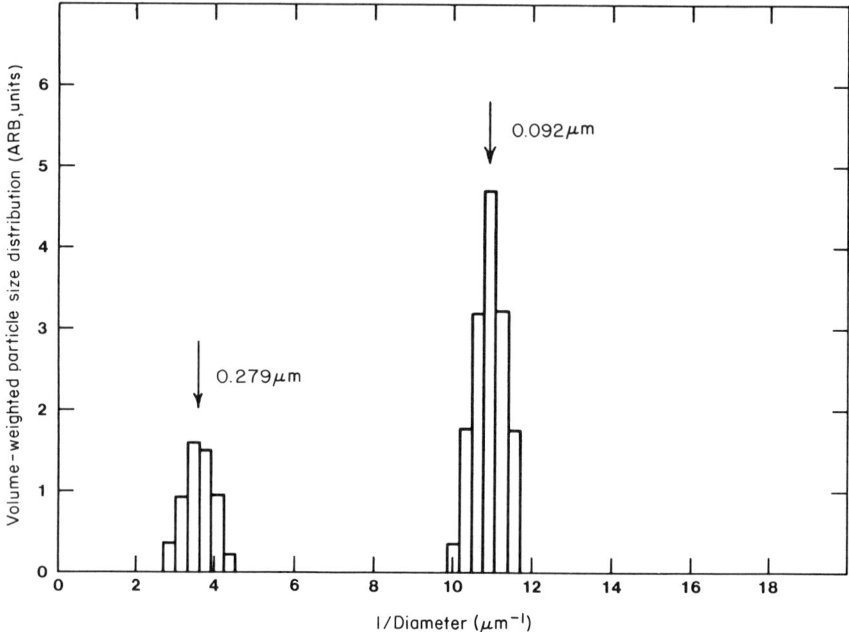

Figure 28. Reciprocal size distribution from photon correlation spectroscopy for a mixture of 0.091 and 0.261 μm polystyrene particles of mass ratio 3 : 1, respectively, obtained using the Nicomp bimodal analysis method. Courtesy of Nicomp Instruments.

particle size analysis. Each of the several phenomena results in a somewhat different definition of particle size. Particle size would be best defined in terms of the response function of the particular instrument exploiting a particular technique. The difficult part of the analysis is the interpretation of the instrument response function, or particle size, in terms of particle dimensions. Often, this interpretation is not required, since the distribution obtained for a particular sample can be compared with a reference or control sample with demonstrated characteristics.

Probably the premier techniques for particle size analysis are those based on particle counting (resistive pulse and optical pulse techniques), since the detailed distribution of individuals or events can be obtained with response proportional to particle volume or cross-sectional area, respectively. The HIAC, utilizing the optical pulse technique, is especially interesting in this category, since its response to particle cross-sectional area should provide a close analogy to particle size distribution obtained by microscopy. The HIAC, unlike microscopy, allows many particles to be sized in a very short time, and also provides distribution of the maximum particle cross section directly. Image analysis by microscopy would be an expensive and difficult task for particles with highly irregular shapes. Instrumentation based on counting techniques would find strong application in "unusual" problems in particle size analysis, where the major part of the sample has the desired size and distribution but product performance is severely jeopardized by the presence of a relatively few oversized particles. Such particles might escape detection by other techniques, contribute to a noisy response (PCS), or be removed by the technique (i.e., filtered out by the packed bed in HDC).

Quite often in particle size analysis, information other than the relative size distribution is required—for example, the total amount (volume, mass, etc.) of disperse phase per unit volume of suspension. Most commercial instruments do not anticipate this requirement, although it could readily be obtained from the integrated response of the detector per unit volume of test dispersion. This is especially true in instruments that have a data analysis system.

A final consideration to bear in mind in selecting an instrument based on a particular phenomenon is the possibility of a complex sample, either by accident or intent. The response obtained from any particular phenomenon will, in general, depend upon characteristics of the particle other than size. For example, particles with identical sizes but different specific gravities will appear to be narrow in size distribution using optical and resistive pulse techniques. However, if sedimentation analysis is used, results based on the differences in specific gravity will be obtained, since the particle response to the gravitational or centrifugal field is determined by the ratio of the particle effective mass and frictional coefficient.

It is hoped that this chapter has made the reader aware of the limitations of instruments for particle size analysis, and that it will serve as a guide in understanding and selecting an appropriate instrument.

REFERENCES

1. T. Allen, *Particle Size Measurement*, 3rd ed., Chapman and Hall, London, 1981, p. 399.

2. R. W. DeBlois and C. P. Bean, *Rev. Sci. Instrum.*, **41**, 909 (1970).

3. R. W. DeBlois, C. P. Bean, and R. K. A. Wesley, *J. Colloid Interface Sci.*, **61**, 323 (1977).

4. R. W. DeBlois, E. E. Uzgiris, D. H. Cluxton, and H. M. Mazzone, *Anal. Biochem.* **90**, 273 (1978).

5. M. M. Tirado and J. Garcia de la Torre, *J. Chem. Phys.*, **71**, 2581 (1979).

6. M. Kerker, *The Scattering of Light and Other Electromagnetic Radiation*, Academic Press, New York, 1969.

7. P. J. Goetz, in A. W. Preece and D. Sabolovic (Eds.), *Cell Electrophoresis: Clinical Application and Methodology*, INSERM Symposium No. 11, Elsevier/North Holland Biomedical Press, Amsterdam, 1979, p. 445.

8. L. G. Bunville, 8th Annual Federation of Analytical Chemistry and Spectroscopy Societies Meeting, Abstract 217, Philadelphia, 1981.

9. L. G. Bunville, unpublished results.

10. R. C. Boltz, Jr. and P. Todd, in P. G. Righetti, C. J. Van Oss, and J. W. Van der Hoff (Eds.), *Electrokinetic Separation Methods*, Elsevier/North Holland Biomedical Press, Amsterdam, 1979, pp. 241–242.

11. D. W. Mason, *Biophys. J.,* **16,** 407 (1976).

12. J. Vinograd, R. Bruner, R. Kent, and J. Weigle, *Proc. Natl. Acad. Sci. U.S.,* **49,** 902 (1963).

13. D. J. Nagy, Ph.D. Dissertation, Lehigh University (1979).

14. H. Small, *J. Colloid Interface Sci.,* **48,** 147 (1974).

15. R. J. Remiarz, J. K. Aggarwal, and E. M. Johnson, *TSI Quarterly,* **8,** 3 (1982). (Available from TSI Inc. See Appendix for complete address.)

16. V. Vouk, Ph.D. Dissertation, London University, 1948.

17. Y. Almog, S. Reich, and M. Levy, *Br. Polym. J.,* **14,** 131 (1982).

18. A. L. Wertheimer, M. N. Trainer, and W. H. Hart, Symposium on Advances in Particle Sampling and Measurement, U.S. Environmental Protection Agency, Research Triangle Park, N.C., 1978.

19. P. J. Wyatt and D. T. Phillips, *J. Colloid Interface Sci.,* **39,** 125 (1972).

20. D. D. Cooke and M. Kerker, *J. Colloid Interface Sci.,* **42,** 150 (1973).

21. P. N. Pusey, D. E. Koppel, D. W. Schaefer, R. D. Camerini-Otero, and S. H. Koenig, *Biochemistry,* **13,** 952 (1974).

22. B. Chu, *Laser Light Scattering*, Academic Press, New York, 1974, pp. 220–222.

23. R. Pecora, *Annu. Rev. Biophys. Bioeng.,* **1,** 257 (1972).

24. S. Broersma, *J. Chem. Phys.,* **32,** 1626 (1960); **74,** 6989 (1981).

25. Ch. Gähwiller, *Powder Technol.,* **25,** 11 (1980).

26. R. W. Lines and B. V. Miller, *Powder Technol.,* **24,** 91 (1979).

27. C. A. Daniels and A. A. Etter, *Powder Technol.,* **34,** 113 (1983).

CHAPTER

2

THE APPLICATION OF
PARTICLE CHARACTERIZATION METHODS
TO SUBMICRON DISPERSIONS AND EMULSIONS

MICHAEL J. GROVES

Pharmaceutics Department, College of Pharmacy,
The University of Illinois at Chicago,
Chicago, Illinois

1. INTRODUCTION

Characterization of the size of component particles in industrial dispersions is often required for research purposes and is a routine and essential part of the overall quality control procedures invariably applied to the products. Characterization methods have been extensively reviewed (1–3), but if the average particle in a dispersed system is at or below the limit of discrimination by the light microscope (about 1 μm), many of the regular sizing methods become inappropriate. Submicron dispersed systems have become technically more important during the past decade in a number of industries, and methodology for sizing has shown a parallel development.

43

The biological activity and effectiveness of pesticides and herbicides are critically affected by the size of the active component used for application. As a result, submicron systems have been developed for commercial use.* In pharmacy, the effect that particle size has on the biological activities of water-insoluble drugs has been well-recognized for many years (4). However, the reduction of solids to submicron-sized powders is expensive and counterproductive. Not only are electrostatic surface forces exposed, which results in the aggregation of the fine particles, but handling such systems also inevitably leads to an increased risk of dust explosions and cross-contamination problems in a factory environment. In addition, there is an increased risk to health when workers are exposed to the possible inhalation and absorption of often potent bioactive materials. In these situations, it is more advantageous to handle the bioactive materials as emulsions or as dispersions in solid–liquid systems. In these forms, manipulation and processing are facilitated.

In the medical field, fine (submicron) emulsions of vegetable oils are now routinely administered to patients as parenteral calorie sources. In these products, it seems likely that administration of sufficient numbers of emulsion droplets close to or greater than the dimensions of a blood capillary could produce undesirable side effects (5). Quality control procedures applied to those systems should ideally determine that the products do not contain appreciable numbers of particles larger than a micron. Intravenous emulsions also offer some promise as drug delivery systems, and the same is true for liposomes or phosphatide vesicles varying in size from 10 to 1000 nm (6). In both cases, it is likely that a critical performance factor is size (7).

Outside the biological and medical areas, the performance of pigment dispersions in terms of covering capacity and color development is also critically controlled by the size of the primary pigment particles and the number of primary particles forming aggregates (8). Although not emulsions in the strict sense of the term, polymer latex emulsions are widely employed as paints. Essential properties, such as covering capacity, penetration, film-forming ability, "brushability," and stability are all critically influenced by the mean size and spread of size around the mean. Frequently, polymer emulsions are in the submicron range and much of the recent progress in size characterization methodology can be directly attributed to developments in this particular industry.

* A list of addresses for manufacturers of commercially available instruments is given in the Appendix.

2. THE MEANING OF THE TERM "SIZE" IN A SUBMICRON DISPERSION

Intuitively, the term "size" in a macroscopic system means the diameter of the average particle, and is often associated with some parameter that provides a description of the width of the size range from the smallest to the largest particle in the system. If the particles can be seen and measured, this concept provides little or no difficulty, and sizing methodology enables the appropriate parameters to be determined.

However, when dealing with submicroscopic or submicron systems, semantic problems arise, because situations occur where a hard and fast definition of the term "particle"—and, therefore, of "size"—becomes somewhat imprecise.

Submicroscopic particles are usually referred to as colloids. A colloid can be defined as a two-phase system with *discrete* particles of the disperse phase that are between 1 and 100 nm (10–1000 Å).

These limits are somewhat imprecise in themselves and are intended to describe particles that are smaller than the lowest level of discrimination by the light microscope, but are capable of detection by the Tyndall beam effect. For example, Faraday prepared colloidal gold solutions that appeared to be clear when placed directly between the observer and a light source. When viewed laterally, however, the light appeared as a yellow track in the solution, and this was interpreted as being due to light scattered from the dispersed gold particles at right angles to the incident beam.

The colloidal size range encompasses macromolecules as well as aggregates of smaller molecules. The surface forces around an aggregated colloidal particle are less per unit mass than the forces around the constituent molecules, if they were dispersed in a "true" solution. The fundamental properties of true and colloidal solutions are the same qualitively, but differ quantitatively, because of the differences in particle size and structure. There is also a range of size in which particles between approximately 0.1 and 2–5 μm have properties between colloids and those of macroscopic dispersions or suspensions. The behavior of colloidal particles is influenced strongly by their environmental conditions. This provides a challenge when attempting to characterize the size of particulate systems with diameters anticipated to be below 5 μm, since properties of the systems can be readily affected by the conditions under which the analysis is carried out.

In addition, it is necessary to closely consider what is actually defined as the "size" of a colloidal particle, especially with particles below 50 nm in diameter. The "particle" itself, for example, might consist of a hydrated protein molecule, a discrete solid particle, or a stabilized droplet of oil. It is preferable in this review to consider particles that have definable, discrete surfaces. The discrete molecular fragment offered by a molecule of, say, hemoglobin will not be considered, but a colloidal gold particle would be. Emulsion droplets (e.g., oil particles dispersed in water) offer larger problems of definition, since they consist in the main of a central hydrophobic "core" and a stabilizing exterior often comprised of a high molecular weight surfactant or dispersed colloidal entity, such as a polysaccharide (tragacanth or acacia gum) or a protein (e.g., casein). Colloidal particles inevitably demonstrate a random movement even under steady state conditions. This is caused by the kinetic energy of the continuous constituent molecules and is known as Brownian movement. By applying a centrifugal field, the particles can be made to move in a consistent direction. Nevertheless, the particles will not move as a discrete, easily definable entity, especially if the surface has a layer of stabilizing counter ions around it. A proportion of the stabilizing layer will move with the surface, establishing a slippage plane, the Stern layer, around the surface. The thickness of the slippage plane is affected by environmental features such as the pH, temperature, and concentration of electrolytes, so that measured size characterization parameters may be influenced by the nature and conditions of a diluent prior to analysis. For particles larger than 50 nm, the thickness of the slippage plane becomes relatively less important.

The stabilizing layer of emulsion particles can itself be influenced considerably by environmental factors, especially if it comprises surface active material. Surface active agents, or surfactants, form molecular associations with water. These associations are defined as states of matter in which the degree of molecular order lies between the almost perfect long-range positional and orientational order found in solid crystals and the statistical long-range disorder found in isotropic amorphous liquids and gases. These mesomorphic states of matter therefore have some of the properties of liquids, in that they may lack rigidity or resistance to deformation, and may also possess optical anistropy, a property associated with solid crystals (9, 10). With surfactant stabilized emulsions, the mesomorphic interfacial layer effectively forms a third phase in the presence of two other phases (11). An emulsion is no longer defined as a dispersion of one liquid phase in another liquid in the form of droplets (12), but rather as liquid droplets and/or mesomorphic systems dispersed in a liquid (13). Although an emulsion assumes two liquid phases dispersed in each other, one of the phases could solidify after the initial emulsification process so that, under ambient conditions, the system is in effect a dispersion, that is, a solid in a liquid or vice versa (14). Since emulsions are inherently thermodynamically unstable, this solidification process is often resorted to as a means of improving long-term stability. Interfacial modification produced by adding absorbed surfactant will assist in minimizing interparticulate interactions by conferring a surface charge that repels particles approaching each other during Brownian movement. An alternative mechanism is to provide a diffuse layer of long chain molecules around the droplet that causes steric hindrance between adjacent particles.

In either case, it becomes increasingly difficult to precisely define the meaning of "particle" and, therefore, of the size that can be measured. In a recent investigation into the properties of self-emulsifying oils, Iranloye (15) was able to distinguish between emulsions consisting of droplets surrounded by a visco-elastic gel network of surfactant phase and those whose droplets were not so protected. Similarly, emulsions stabilized by a steric mechanism involving a long chain surfactant, protein, or polysaccharide system can be considered to have a diffuse interfacial layer. For example, some polymer emulsions have relatively rigid core structures and diffuse interfacial layers and others "soft" cores, but little in the way of interfacial material. As discussed in another section, the passage of these types of emulsion through a porous bed is a direct means of differentiating between these types of emulsion systems.

The "size" of two otherwise identically sized "core" droplets—one with a diffuse interfacial layer and the other without—will differ in a centrifugal sizing method, because the size of the Stern layers differ and the interfacial layer has a different density from the core material, thereby influencing the overall density of the moving particle. The same possibility exists for the light scattering measurement method since, in the case of a diffuse interface, light may not only scatter from the core but also from light scattering centers in the diffuse layer itself, with the net effect that the particle appears to be larger. Obviously, these effects are relatively less important as the particle becomes larger and the thickness of the adsorbed layer becomes proportionally less predominant.

Specimen calculations are usually enough to indicate the influence of these factors on a measured size parameter. Generally, the effect is relatively trivial for particles larger than 200 nm, but may be influential as the size is decreased, depending on the thickness of the interfacial layer. In a system consisting of a spread of sizes around a mean, which is usually the case in the real world, this interfacial effect may not be relevant at the top end of the size range. However, the smaller particles may be affected, because of the likelyhood that the interfacial layer would be of the same thickness and density for all particles in the system, irrespective of their droplet size. The net effect of this influence could be a distortion of the particle size distribution. For example, in a centrifugal analysis, it is quite possible that smaller particles may move less rapidly than predicted by Stokes' equation, although the same law is perfectly valid for the larger particles at the top end of the same distribution. The small particles may, therefore, appear to be smaller than they really are, thereby increasing the apparent spread of the distribution.

3. METHODS OF PRESENTING SIZE DATA

Before discussing the methods of characterizing size, it is necessary to review the definitions of size related parameters. In an effort to define the term "particle," Heywood (16) decided that no upper limit could be placed on the size of a particle owing to its geometrical properties, but a lower limit was clearly defined as a molecule of the substance, since subatomic particles are not under consideration. Heywood suggested that the significant property of a particle is that it is a discrete portion of matter that may be small in relation to the space in which it is dispersed, but that is not necessarily small in absolute size. At its limiting condition, a particle may be defined as a point–source of matter in an unbounded space. This definition refers to a single particle and, in practice, we are generally concerned with assemblies of particles, sometimes having widely different sizes and shapes. From a technical standpoint, we need to define some parameter of a particle within the system that is descriptive of the whole. Although a bottle of an emulsified product may contain as many as 10^{16} particles, the physical properties of that product (such as viscosity or biological activity) may be adequately described in terms of the diameter of a single "average" particle, with a second numerical parameter describing the relative width of the distribution around the mean particle diameter.

In the majority of liquid-in-liquid emulsion systems, the particles are spherical. If all particles were spherical in shape, the only necessary size measurement would be the diameter. In an imperfect world, however, the analyst is often concerned with particles of irregular and uneven shape. Heywood pointed out that irregular particles have only one unique dimension, the maximum separation along a straight line between two points on the surface (17). This is not always easy to measure and does not adequately describe the properties of the average particle. The size of individual particles is usually expressed in terms of the diameter of a circle or a sphere equivalent to the particle with regard to some stated property. This diameter, d, can be defined as follows:

d_a The projected area diameter or the diameter of a circle having the same area as the profile of the particle when viewed in a direction perpendicular to the dimensions breadth (B) and length (L).

d_p The projected perimeter diameter or the diameter of a circle having the same perimeter as the profile of the particle when viewed from above.

d_v The volume diameter or the diameter of a sphere having the same volume as the particle.

d_s The surface diameter or the diameter of a sphere having the same surface area as the particle.

d_{ss} The specific-surface diameter or the diameter of a sphere having the same specific surface as the particle.

d_f The free-falling diameter or the diameter of a sphere having the same free-falling velocity as the particle. (If the diameter-to-velocity relationship is in accordance with that predicted by application of Stokes' law, the symbol d_{st} is used.)

Other dimensions for describing irregularly shaped particles include the equivalent size of the smallest square or round aperture through which a particle will pass. These dimensions represent other properties of the particles and differ considerably from each other in numerical magnitude, especially if the particles are flakes or are needle-shaped.

Particles in a system are not always the same size or shape, and a spread around the mean should be anticipated. To describe the spread, it is necessary to make some assumptions about the distribution. It may be sufficient to describe the upper and lower limits of the distributions as d_{min} and d_{max}, or as the upper and lower quartiles. Only rarely is a symmetrical distribution around a mean encountered in practice; often, there are many more particles smaller than some defined particle with properties that adequately describe a system's overall properties. A form of exponential relationship is usually adequate to describe the properties of the system as a whole, and the logarithmic–probability relationship is a convenient approximation. This law allows an adequate definition of a mean size, and the spread is described as the width of one standard deviation around the mean (i.e., the ratio $d_{50\%}:d_{84\%}$ or $d_{16\%}:d_{50\%}$). By convention, this standard deviation is always more than unity, a value of 1.0 being a truly monodisperse system. One advantage of this definition of size spread is that the averages, as previously defined, may be numerically related to each other—the Hatch–Choate conversion (17). If a system distribution is described by the logarithmic-probability law, the results obtained by one method of size analysis may be related to those obtained from another. Logarithmic-probability distributions tend to fit materials prepared by homogenization or fracturing.

4. DIRECT METHODS OF CHARACTERIZATION—MICROSCOPY

The only direct or absolute method of size characterization is the optical or light microscope, where the operator is involved in observing the particles, making judgments, and taking measurements of their properties.

The routine use of a light microscope is almost mandatory in any laboratory where there is an interest in particle characterization, although use of the instrument may be confined to qualitatively examining the state of dispersion, without necessarily making measurements. Even under optimum viewing conditions, with adequate separation and contrast between the particles, a realistic lower limit for diameter measurement by light microscopy is around 2 μm. It is possible that this lower limit could be reduced by the use of differential interference contrast microscopy, but even in this case, claims to any accuracy at sizes below 1 μm should be treated with some caution. Generally speaking, microscopy is invaluable for making subjective impressions but is tedious and imperfect for making objective measurements. The accuracy of the measurement process has been considerably increased in recent years by the use of instruments employing image shearing techniques, for example, the Coulter "Shearicon," the Fleming Instruments microscope, and shearing eyepiece devices such as that made by Vickers. These devices function by shearing an image optically or mechanically until the two images just touch—the amount of shear required is a measure of the size of the object.

Unfortunately, the reduced definition of the light microscope as the sample diameters approach the dimensions of the illuminating light, combined with the increasingly violent Brownian movement as the particles become smaller, tend to make direct microscope observation of limited value for the characterization of colloidal materials.

5. INDIRECT MICROSCOPY—IMAGE ANALYSIS AND ELECTRON MICROSCOPY

Automated methods for image analysis have become very sophisticated in recent years and may be applied to direct images in a microscope or to photographs obtained indirectly, using an electron microscope. The image is scanned by a high resolution plumbicon or vidicon tube to convert the optical image into a video signal. After this stage, the manipulation of the signal is only limited by the capacity of the computer or the imagination of the operator. When a large number of images are analyzed routinely, the high cost of the instrumentation is more than offset by the increased accuracy and workload. Dedicated computers attached to the instrument are capable of printing out features such as the number of particles of a predetermined size within a given portion of a field, the chord analysis of the dispersion, or, in a mixed system, the relative properties of the components. A number of instruments are available today, including the original Quantimet device, which has a wide variety of user options.

Attention should be drawn, however, the danger of becoming overconfident in data manipulation. An improved degree of sophistication does not necessarily improve the basic quality of the original information being manipulated. This comment, of course, applies to all types of instrumentation. In the case of image analyzing instruments, all of the disadvantages, such as insufficient discrimination or poor dispersion of the original image-making technique, are retained. In addition, the method depends on the degree of precision with which the magnification of the original image is known in the first place (3).

Electron microscopy, used in either the transmission or the scanning (surface) mode, is becoming more widely available as the cost of equipment is decreasing. The method remains the only means available for the visualization, sizing, and characterization of many submicron particulate dispersions. However, in practical terms, mounting and staining remain crucial to the examination and, more to the point, interpretation, of dispersions. The use of an electron microscope becomes limited by the operator's skill. Staining agents include osmium and molybdenum salts. The method requires a high vacuum, so that artifacts may be produced in a preparation that places limitations on the interpretation of data. The vacuum conditions may result in the shrinking of specimens, and until recently, preparations containing water could not be examined. The introduction of the freeze fracturing technique has allowed replicas to be made of dispersions in water. This method is unique in the field of particle characterization, since the dispersion usually need not be diluted. The intact material, therefore, can be examined without risking untoward effects caused by the dilution process and the nature of the diluent. Artifacts may occur on occasion, when the advancing ice front squeezes particles together and increases the local electrolyte concentration, but one can guard against this effect.

Freeze fracturing as a technique has been invaluable in many areas. For example, earlier postulates about the nature and dimensions of in-

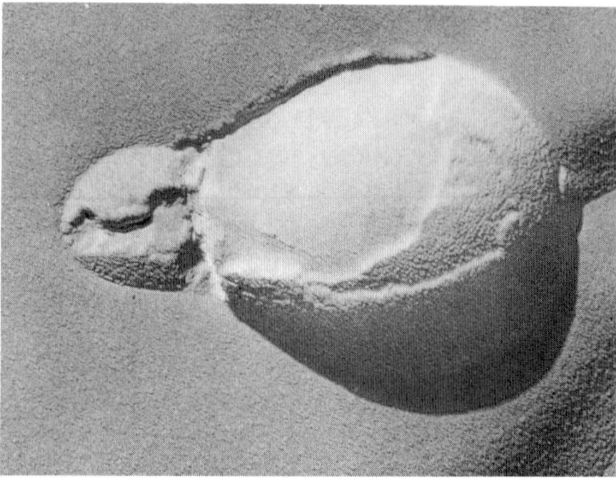

Figure 1. Scanning electron photomicrograph of a freeze-fractured sample of triglyceride droplets stabilized with phospholipid, showing the interfacial structures that suggest multilayers around the droplets. The fractured surface is lightly shadowed with gold.

terfacial layers around emulsion droplets have been substantially confirmed by direct visualization (Figure 1).

It is only rarely possible or necessary to work directly from an electron microscope screen, and photographs are always made of any one field. Since it is necessary to examine a large number of fields, automatic or semiautomatic scanning devices are advantageous to generate sufficiently useful data from an electron microscope facility. These are often not available, so that interpretation of data is either carried out manually or on a qualitative level. Unfortunately, to approach statistical accuracy when characterizing the size of particulate systems, a sufficiently large number of different fields (i.e., samples), must be taken. Because a sufficient number of measurements is required, it is tempting to be less than thorough. Paradoxically, sampling error becomes much higher as the particle diameters increase toward the realistic upper limit of the method, ~ 1 μm, because the size of the field is proportionally smaller and a large number of fields are examined.

Another problem concerns estimates of size. The size of the image is dependent initially upon the voltage across the electron source, and there may be several stages of subsequent image magnification, causing additional sources of error. This overall error factor is partially overcome by calibrating the system with relatively monosized polymer latex particles or by diffraction gratings. However, the latex standards are themselves often sized by electron microscopy. Caution is therefore required in the interpretation of much of the published data on electron microscope sizing, especially if the authors do not provide details of calibration or the number of samples taken.

Scanning electron microscopy is proving to be a useful analytical tool. Surfaces, when bombarded by electrons, emit secondary electrons and x-rays that are characteristic of the surface. It has become feasible to unambiguously identify the elemental composition of individual particles, provided that the atomic number is greater than 14. The technique also makes it possible to derive particle mass, maximum thickness, and average thickness for single particles, so that characterization of particulates has become more intensive and valuable. Advances in this area can be anticipated.

6. HYDRODYNAMIC CHROMATOGRAPHY

Hydrodynamic chromatography (HDC) is the technique developed initially for the characterization of polymer latices by Small (19) (see Chapter 9 for applications to high molecular weight polymers). Column chromatography is used to separate particles according to size. At the time the concept was originally tested, the chromatography was mainly confined to the separation of materials in a column packed with solid particles; an appropriate detection system was used at the end of the column (see Chapter 8). Small attempted to determine whether size fractionation could be obtained, based on van der Waals interaction between the packing surface and a "solution" phase consisting of submicron particles in suspension. It was rationalized that the larger particles should be retained longer than the smaller, thereby affecting a size-based separation. Initial experimentation demonstrated that separation occurred, but in the direction opposite to the prediction. This prompted a closer investigation of the phenomenon and led to the development of what is now a potentially useful and promising technique. The basic equipment consists of a chromatographic column packed with beads. An aqueous solution, the mobile phase, is pumped through the column at a known flow rate. The nature of both the column packing and the mobile phase are critical to the separation. A pulse of a colloidal suspension is injected into the column in just the same fashion as any other chromatographic separation technique. Mobile phase from the end of the column is passed through a suitable detection system, such as a flow-through spectrophotometer of the type used in liquid chromatography, and the detector response of the colloid is determined as a function of elution time (see Chapter 8).

The rate of elution of a fraction from a column is expressed by a dimensionless parameter, the retention factor, R_f. This number is the ratio of the rate of migration of the particulate peak to the rate of the mobile phase flow through the interstitial volume of the column and is independent of the flow rate or column dispersion. The interstitial flow rate can be determined by the use of suitable markers, such as sodium dichromate, injected into the column with the suspension.

Experimentally, it was shown (20) that the R_f increased with increased particle diameter, and separation improved as the size of the packing material was decreased. The elution volume of a sample was not affected to any marked degree by the composition of the packing material itself.

Addition of electrolytes to the mobile phase slowed the rate of elution to the point where the elution order of a mixture was reversed, larger particles eluting more slowly than the smaller. This was accounted for by suggesting that electrolytes suppressed the electrostatic repulsion existing between packing surface and particle. At low electrolyte concentrations, this repulsion was sufficient to cause the smaller particles to move into the higher velocity streamlines located in the center of the interstitial spaces.

The R_f is always greater than unity, suggesting that the particles themselves are moving faster than the mobile phase. This is explained by a hydrodynamic effect. The particles are assumed to be rigid, so that they cannot approach the wall of the interstitial space any closer than the particle radius. Since the mobile phase velocity is higher in the center of the interstitial space, there is a tendency for particles to be forced toward the center. The resultant particle velocity is faster than the average mobile phase velocity. This hydrodynamic effect qualitatively explains why larger particles move faster than smaller ones, but is not entirely satisfactory, since subsequent work has demonstrated the importance of electrostatic and van der Waals' surface forces.

The complexity of the factors involved in separating colloid particles in a packed column has raised some doubt about the use of calibration of a hydrodynamic chromatographic system. Initial data (20) suggest that, if the same conditions were maintained exactly, particles of one chemical species could be used to calibrate a column for particles of another species. Reasonable agreement has been obtained between different methods (21).

Because particles eluting from a column are detected by optical spectroscopic methods and the particles are generally smaller than the wavelength of the illuminating light, it is necessary to correct the experimental chromatograms for the dependence of turbidity on particle size as well as dispersion (22), as discussed in Chapter 8.

In general, HDC has been successfully applied to the size characterization of a number of polymer latices. The method is applicable to the size separation of particles between 20 nm and 1 μm, if they are rigid. Particle density does not affect separation, but deformability can do so. The application of HDC to liquid–liquid emulsions has not been successful.

The method has the advantage that it can be carried out using commercial column chromatographic equipment, with readily available columns and column packing materials. If mobile phases are filtered to minimize microbiol growth in the column packings, or contain a preservative such as sodium azide, the columns will function without trouble for several months.

Elution times depend on flow rate and column length. Under optimum conditions, samples may be eluted completely in 10 min or less, a number of samples being run in sequence. This enables 30–40 samples to be run in a routine working day, which is an enormous advantage from the point of view of quality control. HDC may be capable of improved flexibility and application to situations outside those originally envisioned.

7. NEPHELOMETRIC AND LIGHT SCATTERING METHODS

Indirect methods of size characterization that use radiation scattering, especially those involving alteration or scattering of light, have a number of advantages. In particular, these methods may be fast, relatively inexpensive, and have practically no lower particle size limit. These techniques require small amounts of material and, most importantly, do not induce artifacts in the sample itself, which can lead to misintepretation of the results. Also, a property of the system is that it effectively monitors samples, rather than accumulating data from individual measurements of individual particles, so that "average" size or size characteristic may have a more adequate meaning. However, until recently, it was not feasible to measure the spread of the distribution around the mean. With the advent of laser Doppler techniques, this problem has become less of an issue.

Scattered radiation may be light or x-ray. Unfortunately, as the size of the dispersed particles approaches the dimension of the wavelength of the incident radiation, the classical laws of geometric optics become invalid. The advantage of x-rays is that they have a much shorter wavelength, which allows the application of optical laws. The disadvantage is that x-radiation passes through materials with atomic weights below about 14, thereby rendering organic materials containing substantial amounts of carbon, hydrogen, and nitrogen transparent. For this reason, scattering of light at wavelength dimensions appreciably larger than the particle diameter has received attention and is better understood (23).

Lord Rayleigh (24–26) investigated the scattering of light by small particles and summarized his findings in the following equation. The intensity of light scattered at angle θ is

$$I_s(\theta) = \frac{9\pi^2 N v^2}{2d^2\lambda^4} \cdot \frac{(\eta_i - \eta_{ii})}{(\eta_i^2 + 2\eta_{ii}^2)} \cdot (1 + \cos^2\theta)\, I_0 \qquad (1)$$

where N is the number of particles in a volume v, d is the distance between the measuring unit and the sample, λ is the wavelength of the incident light of intensity I_0, and η_i and η_{ii} are the refractive indexes of the disperse phase and the continuous phase, respectively. A major restriction of Equation 1 is that it is not valid for particles much larger than approximately one-tenth of the dimension of the incident radiation wavelength. This restriction assures that each particle has a uniform internal field induced in it when the incident wave passes through it. The equation does not take into account the amplitude of light refracted two or more times within the same particle. Since only single refractions are considered, this limits the mathematical analysis of the system to those situations where the refractive indexes of the disperse and continuous phases are close to each other (23).

Van der Waarden (27) adapted the Rayleigh equation to calculate the mean particle diameter in emulsion systems by assuming that the emulsion droplets were monodisperse in size and then applying a correction factor for multiple scattering. When neither of these assumptions can be justified, calculation of the light scattering parameters becomes complicated.

Mie (28) formulated equations to account for the complete light scattering pattern in a particulate system by considering the electrical field inside and outside of a single particle. The scattering intensity was calculated as a series of expressions, each a function of various scattering angles, the difference in refractive index of the particle sphere and the continuum, the wavelength of the incident radiation, and the diameter of the sphere. This approach is complex and only became realistic with the advent of the computer. Originally, the theoretical approach referred to simple spherical particles of the same size in a simplified model system, but with increasing sophistication of computer technology, realistic systems have been examined. Applications of this approach were described by Robillard and Patitsas (29–32) and, in some cases, a degree of polydispersity around the mean may be derived.

The complexity of the light scattering pattern resulted in the development of at least one instrument that may be used to measure this pattern

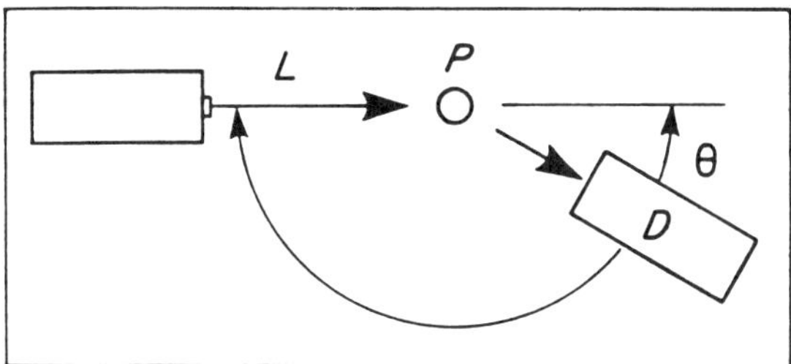

Figure 2.   Diagram of Science–Spectrum light scattering photometer where L = laser beam, P = single particle, D = detector, and θ = scattering angle. Reproduced with permission from Wyatt Technology Co.

from a single particle, the Science Spectrum Differential® II light scattering photometer manufactured by Wyatt Technology Company. This instrument contains electrical means for the capture and indefinite levitation of a single microparticle in a laser beam while a rotating photodetector allows the intensity of scattered light to be measured as a function of the angle of detection (Figure 2).

The so-called differential light scattering (DLS) patterns may also be measured from suspensions of particles, using a Differential® I system wherein the levitation module is replaced by a simple structure that holds a cuvette containing the sample. The laser beam (vertically polarized) passes through the sample, which is again scanned by a photodetector. The effects on such patterns of changing size distributions for suspensions containing particles of the same average size are illustrated in Figure 3. From these recordings, the mean size and size distribution may often be obtained by matching known and unknown samples (33). During the past decade, an atlas of characteristic spectra has been generated for empirical matching, though computer inversions may also be used. This approach is a semiempirical one and requires some *a priori* knowledge of the particles present, but can often yield important particulate parameters, including identification, mean size, size distribution, and the refractive index and structure features. Because the instrument can be fitted with lasers of different wavelengths, including helium–neon, xenon ion, helium–cadmium, and a tunable argon–ion source, the potential applications are numerous. In practice, the instruments have been used to study bacteria, polymers, clays, beverages, colloids, pigments, and most liquid-borne particulate ensembles. A modular modification of the instrument permits the measurement of single particle scattering properties under a variety of physical conditions (34).

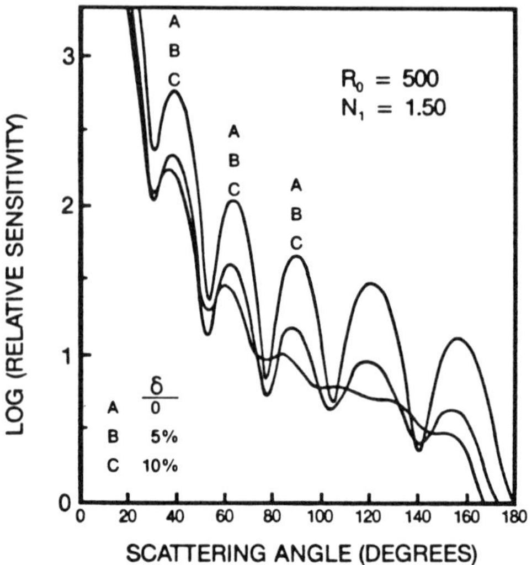

Figure 3. A sample page from the Science–Spectrum Atlas showing the light-scattering curves of various particle distributions in aqueous suspensions. These particles have a mean radius of 500 nm with an index of refraction of 1.50; δ is the half width of the distribution at half maximum, represented as a percentage of the mean size (in this case, 500 nm). Reproduced with permission from Wyatt Technology Co.

Meehan and Beattie (35) described a method of estimating a size distribution by measuring the turbidity of a suspension at two wavelengths. More recently, Yang and Hogg (36) improved the method by using a recording spectrophotometer at four wavelengths.

The turbidity, τ, of a suspension of particles is a measure of the reduction in intensity, caused by light scattering, of the transmitted beam and is defined by:

$$\tau = \frac{1}{l} \ln \frac{I_0}{I} \tag{2}$$

where l is the length of the light path in the suspension and I_0 and I are the intensities of the incident and transmitted beams, respectively.

If the laws of geometric optics were obeyed and the suspension consisted of identically sized spherical particles, the turbidity would be given by:

$$\tau = v\, \pi r^2 K_t \tag{3}$$

where v is the number concentration of particles in the suspension, r is the radius of the particles, and K_t is the total scattering coefficient, defined as the ratio of scattering cross section to the geometric cross section. In general, the values of K_t require that it be evaluated from the Mie theory (23, 37), but this is not always feasible. This equation only accounts for the light scattered exactly in the forward direction. However, in any practical instrument, the measured intensity is made up of contributions from the transmitted light plus any light scattered in the forward direction in a cone of solid angle ω determined by the geometry of the optical system. Equation 3 assumes ω to approximate zero. By including a correction factor, R, which is equal to K_w/K_t, where K_w is the effective scattering coefficient, it is now possible to write:

$$\tau = RK_t\pi r^2 v \tag{4}$$

In a practical system, the particles are not all identical and Equation 4 must be written in the form:

$$\tau = \pi \int_0^\infty RK_t r^2 d_v(r) \tag{5}$$

where $d_v(r)$ is the number of particles per unit volume of suspension whose radius lies between r and $r + dr$. If the particle size distribution is described in terms of a density function $n(r)\ dr$, the mass fraction of particles whose radius lies between r and $r + dr$, Equation 5 can be rewritten as

$$\frac{\tau}{C} = \frac{3}{4}\rho \int_0^\infty R(r)K_t(r)r^{-1}n(r)\ dr \tag{6}$$

where τ/C is the specific turbidity of the suspension and C is the concentration of particles of density ρ.

Numerical integration of Equation 6 involving the log-normal distribution was carried out by computer. The method was applied to a number of submicron materials in suspension, with excellent agreement with data obtained by a centrifugal sedimentation method. An advantage of this technique is that it only requires a spectrophotometer by way of equipment, and, once the appropriate computer software is in place, it can be applied to the routine size analysis of submicron materials.

Another less sophisticated approach to the size analysis of a submicron dispersion was provided by Lee and Groves (38), who adapted a commercially available laser nephelometer for this purpose. The instrument, the Hyland PDQ Laser Nephelometer (Figure 4), was developed for de-

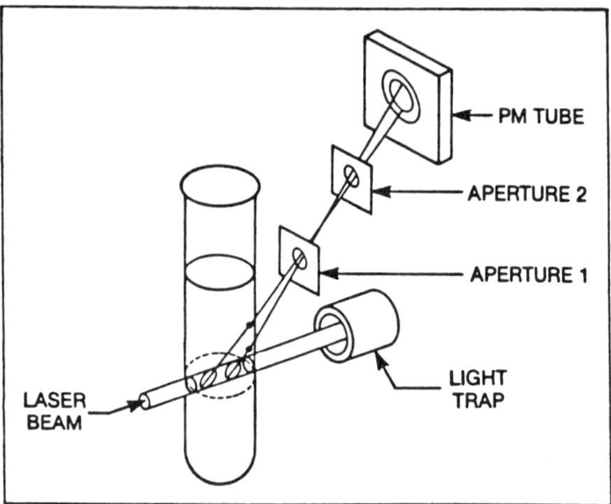

Figure 4.   The Hyland Laser Nephelometer PDQ instrument. Reproduced with permission from Cooper Laboratories Inc., Palo Alto, CA.

termining immunoglobulins by detecting light scattered at 31° in the forward direction from the center of a small cuvette. A memory circuit built into the device allows blank readings to be made of background reagents. Because a polarized light source is used, the Rayleigh equation (Equation 1) requires modification, but many of the factors in the equation may be regarded as constant. These include the distance d, the light wavelength λ, and the angle θ. The value of I_0 is not known, since the ratio of incident to scattered light cannot be determined. Accordingly, the approach of Van der Waarden (27) is not appropriate. The measured intensity of the scattered light may be represented by:

$$I_s(\theta) = f[nvf(\eta_i - \eta_{ii})] \tag{7}$$

The number of particles, n, in the volume, v, is a function of the disperse phase volume (θ) or the concentration term of the original emulsion. The factor $f(\eta_i - \eta_{ii})$ is a function of the relationship between the refractive index of the disperse phase (η_i) and the continuum (η_{ii}), and the intensity of the incident light, I_0. The intensity of light scattered by a series of particles measured by the nephelometer may be simplified to:

$$J_z = \frac{I_s\theta}{(\eta_i - \eta_{ii})\phi}$$

or $\tag{8}$

$$J_z = f(\text{mean particle size})$$

where J_z is the instrument response factor. All other factors that affect the intensity of the scattered light are constant, because of the instrument design constraints.

Calibration of the instrument suggested that the response was substantially linear below approximately 600 nm, which is presumably the wavelength of the illuminating helium–neon laser (632.8 nm). Particles down to 10 nm could be detected. It was assumed that, if an emulsion system contained particles below 600 nm in diameter, it could be characterized by calibrating the instrument with two emulsions of similar chemical identity, but with differing and known mean diameters. This approach works very well in a production environment where rapid results are required to determine, for example, how many passes are required through a high pressure homogenizer. The method, by no means absolute, depends for accuracy on the precision with which the size of the calibration material is known, and is seriously affected by the presence of large ($> 1\mu m$) particles in the system.

A similar pragmatic approach would appear to be the underlying basis for the simplified spectrophotometer technique recently introduced as the Gaulin Emulsion Quality Analyzer. A diluted emulsion sample is placed in the device and an "absorbance index" number is recorded. This reading is then compared with a predetermined butterfat or oil content of the sample to provide an "average particle size diameter" as a measure of emulsion "quality."

Light obscuration instruments, such as the Royco/HIAC devices, are not discussed in this chapter, since it is generally recognized that the effective lower limit of detection of these devices is around 1 μm (See Chapter 1). Nevertheless, one instrument operating on the light scattering/ obscuration principal, the Climet (Figure 5), is included, because it is claimed to be sensitive to particles in the submicron region. The Climet employs white light illumination and obtains an enhanced sensitivity by using an elliptical mirror optical detection system, 400 nm particles being detectable in a liquid-borne system.

Although the use of a relatively unsophisticated white light source and simple detection system is attractive, the sensitivity of a laser light-scattering device is somewhat higher. There have been a number of laser activated devices for detecting particles in flowing streams. Lieberman (39) recently described a commercial device, the Royco Model 346, which uses a laser forward scattering principle (Figure 6). This device is claimed to detect submicron particles in streams flowing at rates of up to 100 ml/ min. Discriminating between the background scattered radiation and the signal caused by the particles passing through the sensing zone is a problem encountered with most equipment of this type. The Model 346 is

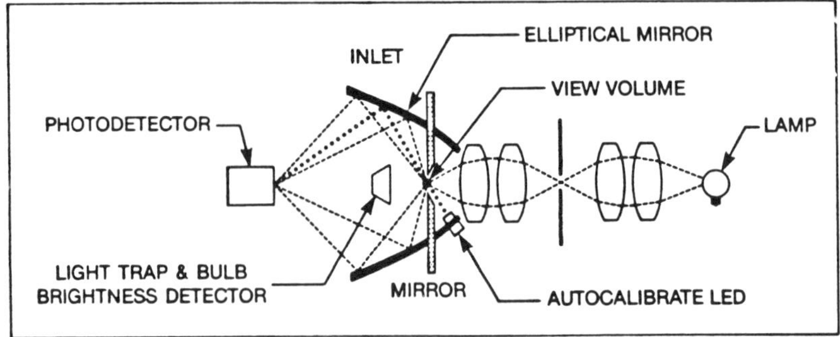

Figure 5. Diagram of the Climet Liquidborne Particle Counter. Reproduced with the permission of Climet Instruments.

apparently capable of producing a 2-to-1 signal-to-noise ratio at 0.5 μm. This is more than adequate to allow detection of particles at this size level and, more to the point, to allow discrimination between sizes in this range.

Instruments based on detecting the diffraction patterns produced when a beam of light is passed through a flowing stream of particles have recently appeared on the market. The theory of these devices was discussed by Talbot (40, 41), and they are generally more suitable for particulate systems with sizes above 1 μm. The Cilas Granulometer 715 (Figure 7) detects the distribution of the light pattern in the focal plane with an array

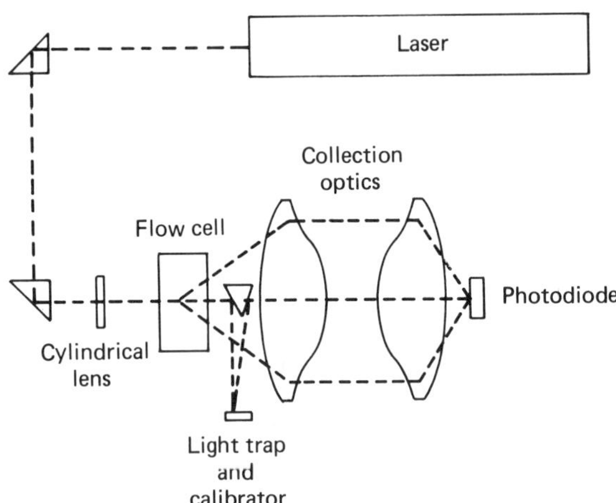

Figure 6. Royco Model 346 Optics. Reproduced by permission of HIAC/Royco Instruments, Division of Pacific Scientific Inc., Menlo Park, CA.

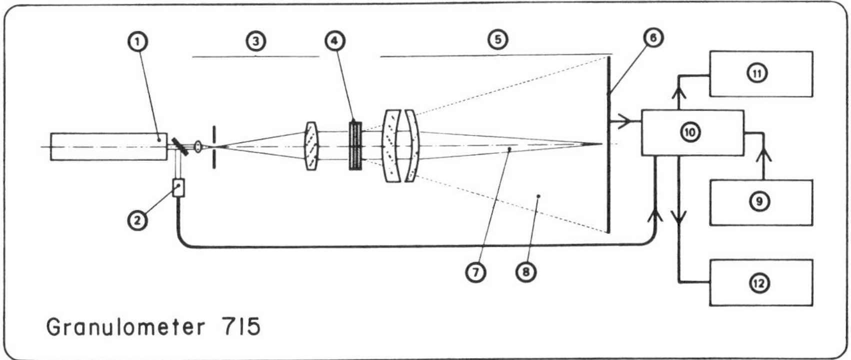

Figure 7. Cilas Model 175 Granulometer: (1) He–Ne laser, (2) reference photodetector, (3) beam expander, (4) sample holder, (5) diffraction pattern analyzer, (6) multicell detector, (7) direct beam, (8) diffracted beam, (9) control keyboard, (10) computer, (11) display unit, and (12) printer. Reproduced by permission of the manufacturer.

of 15 silicon photodiodes. With a laser light source, the instrument has one channel optimistically described as covering the range from 0 to 1 μm although, more realistically, the lower limit of detection is probably around 0.5 μm. Similar devices using a laser to illuminate a flowing stream are the Malvern Autosizer and Leeds and Northrup's Microtrac device, as discussed in detail in Chapters 5 and 6. This latter instrument functions as an on-line analyzer and employs Fraunhofer diffraction pattern masks. In its original form, the Microtrac was not a submicron instrument; but a more recent modification, the Microtrac 7991-3, incorporates an additional tungsten–halogen light source, which detects light of different wavelengths scattered at 90° to the cell. This modification is claimed to extend the sizing range down to 0.1 μm (42). This is illustrated in its white light form in Figure 8. These two instruments are sophisticated on-stream devices that have considerable potential application in a production environment in a quality control capacity. Both incorporate novel and useful data processing software using microprocessors. Calibration for these devices is unnecessary, because they are based on the diffraction principle. These statements by the manufacturers should, however, be regarded with some caution.

8. PHOTON CORRELATION SPECTROSCOPY AND LASER DOPPLER ANEMOMETER METHODS

The invention of the laser in 1961 provided a source of coherent monochromatic light of high intensity, and this resulted in the development of

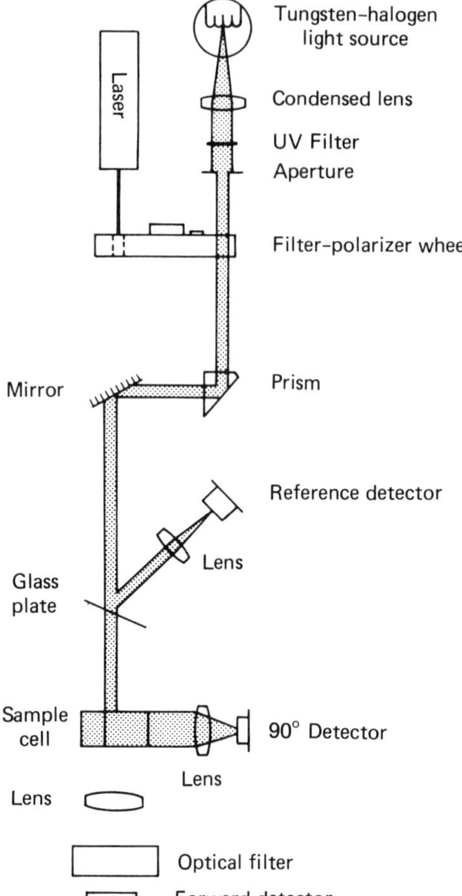

Figure 8. Fine particle configuration for Leeds and Northrup Microtrac 7991-3. Reproduced by permission of the MICROTRAC Products Division of Leeds and Northrup Co.

dynamic light scattering techniques. A suspension of particles within a given space can be regarded as a three-dimensional array. Because the light is coherent, the phase relationships in a beam are maintained, and a random diffraction pattern is formed from the array as it is illuminated with the beam. This random or sparkle pattern consists of alternate bright or dark spots as constructive or destructive interference occurs. The particles in the array are continually moving by random Brownian motion, so fluctuations in the scattered light contain information about the particle size of the particulate system. Dark and light moving diffraction patterns

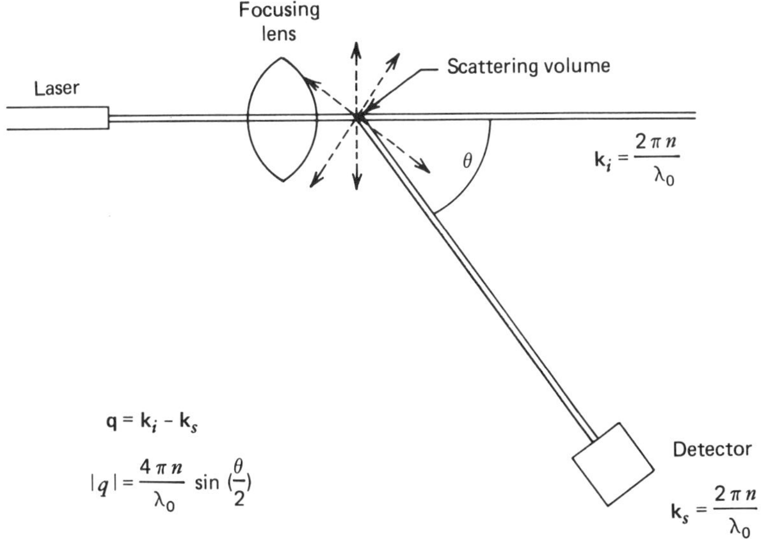

Figure 9. Configuration of laser light scattering photon correlation spectrophotometer. q is the waver vector, n is the refractive index of the suspending medium, k_i and k_s are the incident and scattering wave vectors, respectively, and λ_0 is the wave length of light in vacuum.

caused by the particles result in a time-modulated signal processed by a correlation computer.

The theory of this approach was derived by Pusey and others (43–45) as an outcome of earlier work on radar scattering from distant objects.

A typical experimental setup is shown in Figure 9. Vertically polarized laser light is scattered from a suspension and a photomultiplier detects single photons scattered in the horizontal plane at an angle θ from the incident beam. The signal-to-noise ratio is experimentally increased to a maximum by focusing the beam to increase the coherence area and by having the receiving optics detect no more than a few of these areas.

The scattering particles are effectively diffusing around their equilibrium positions, causing a fluctuation over time of the number of scattering centers in the scattering volume that are seen by the photodetector. The function of the spectrometer is to analyze the intensity fluctuations to obtain the diffusion coefficient. The most efficient way to do this is to average the product of the signal from the photomultiplier and a delayed version of the signal as a function of the delay time, t. This is known as autocorrelation and provides an autocorrelation function related to the diffusion coefficient.

If the sample is truly monodisperse (i.e., all the particles are identical in size and shape), the measured autocorrelation function is given as:

$$C(t) = B + A \exp(-2\,Gt) \tag{9}$$

where A is a constant depending on the optical design and B is a baseline signal determined from monitoring channels on the correlator. The function G is the Rayleigh line width obtained from a spectrum analyzer or autocorrelator and is related to the diffusion coefficient D and the wave vector q by:

$$G = Dq^2$$

where

$$q = \frac{4\pi n}{\lambda_0} \sin\frac{\theta}{2} \tag{10}$$

n is the refraction index of the medium in which the particles are moving and λ_0 is the wavelength of the laser light in vacuum. The diameter of a spherical particle, d, can then be determined from the diffusion coefficient by means of the Stokes–Einstein equation:

$$D = \frac{kT}{3\pi\eta d} \tag{11}$$

where k is the Boltzmann constant, T is the absolute temperature and η is the viscosity of the continuum.

Good agreement has been obtained between photon correlation spectroscopy (PCS) and electron microscopy with latex spheres as relatively monodisperse systems (46). Unfortunately, the analyst is rarely presented with truly monodisperse systems. In a polydisperse system, the light measured is effectively the sum of all the individual correlation functions, and this makes the interpretation of the data more difficult. The method of cumulants is used to deal with continuous size distributions. Here the logarithm of the measured correlation function is fitted to a polynomial in delay time. The coefficient of the linear term is $\bar{G} = \bar{D}\,q^2$, where the bar indicates an averaging process. Rayleigh scattering results if the particles are much smaller than the wavelength of the illuminating light. If the measurements are extrapolated to low angles, the resultant average is the so-called z-average, the next highest average above the weight average. Accordingly:

$$\bar{D}_z = \frac{kT}{3\pi\eta}\left(\frac{1}{d}\right)_z \tag{12}$$

The variance of the z-average diffusion coefficient is twice the coefficient of the quadratic term, that is, $m_z = \overline{D_z^2} - \bar{D}_z^2$. This term is related to the width of the distribution. The relative width of the diffusion coefficient distribution is equal to the square root of Q, where:

$$Q = \frac{m_z}{\bar{G}^2} = \frac{\overline{D_z^2} - \bar{D}_z^2}{\bar{D}_z^2} \tag{13}$$

For spheres, this may be rewritten as:

$$Q = \frac{\left(\overline{(1/(d)_z^2} - \overline{(1/d)_z)}\right)^2}{\overline{(1 - d)_z^2}} \tag{14}$$

This type of analysis yields two moments of the size distribution, namely, the z-average of the inverse of the diameter and its variance. However, as noted earlier, it would be more useful to describe the size distribution by a weight (or volume) average. To do this, it is necessary to assume an appropriate size distribution model, such as a log-normal. The PCS is unable to distinguish the correct underlying model, and the operator has to make the appropriate decision.

In the case of a log-normal distribution with a mass median diameter of d_g and a geometric standard deviation of σ_g, the relationships of Q and the z-average of $1/d$ are:

$$\sigma_g = \exp[\ln(1 + Q)^{1/2}] \tag{15}$$

$$d_g = \left(\overline{\frac{1}{d_z}}\right)^{-1} \exp\, -2.5\ln^2\sigma_g \tag{16}$$

The weight average distribution is $d_g \exp \frac{1}{2}(\ln^2 \sigma_g)$.

Transformation from the simple z-average diffusion coefficient distribution to any other required distribution is relatively straightforward with the appropriate computer processing equipment.

The equipment used will detect particles down to approximately 2 nm and up to approximately 5 μm. The simpler devices utilizing a 90° scat-

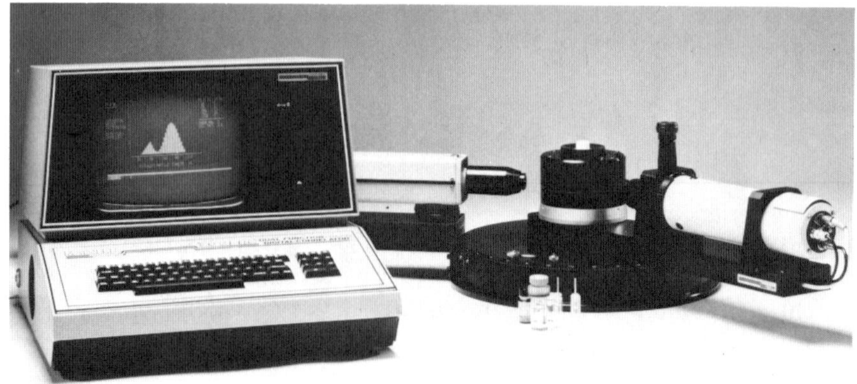

Figure 10.   Malvern System 4600. Photograph courtesy of Malvern Instruments Ltd.

tering angle are valuable if the system is relatively monodisperse and, preferably, scattering in the Rayleigh region in sizes with diameters well below the wavelength of the incident light. The addition of a multiangle goniometer and irradiation with light from lasers of different wavelength increase the flexibility of the PCS technique. This is especially true if the sample has a relatively wide size distribution.

There are a number of commercially available photon correlation spectrometers (see Chapters 1 and 3). The originator was Malvern Company, who markets several versions of this equipment. The Malvern System 4600 instrument (Figure 10) features a variable angle spectrometer, incorporates a self-contained microcomputer, and has a Peltier-effect temperature control system for the sample cell. Construction of this instrument is modular, allowing flexibility in operation, and there is spare capacity   in the microcomputer so that it can serve as a general laboratory computer. Malvern also makes a less expensive instrument, the "Autosizer" (Figure 11). This is a fixed-angle instrument that incorporates a cell temperature control system and an interactive keyboard and visual display unit, with connections to a hard copy printer if required. A somewhat similar instrument has been introduced by Nicomp Instruments, the Model HN5-90 spectrometer with 90° detection fitted with a Model TC-100 computing autocorrelator. An additional source of this type of instrument is Brookhaven.

Coulter Electronics, Inc. has a Model N4 Sub-Micron Particle Sizer shown in Figure 12, which is a successor to the discontinued Nanosizer

Figure 11.   Malvern Autosizer. Photograph courtesy of Malvern Instruments Ltd.

and is microprocessor controlled. The Nanosizer was described by Gah-willer (47) and the schematic diagram is shown in Figure 13 (see Chapter 1). Results from the Nanosizer are displayed as a mean size and a polydispersity index from 0 to 9. The physical significance of the polydispersity index is not entirely clear, but has been discussed by Daniels and Etter (48). These authors found that their results, in particular the polydispersity index, were significantly affected by the presence of a relatively small number of large particles in a system. This observation is likely to be true for all PCS measurement systems. Other comparative studies involving the Coulter Nanosizer include the papers by Lines and Miller (49) and Darling and Shaw (50). The latter group modified their instrument by adding a thermostatically controlled temperature cell holder. The newer Coulter N4 instrument incorporates a Peltier-effect thermostat and provides a number of options for displaying information.

The Malvern Particle Charge instrument (Figure 14) is based on a similar principle, but with a different optical arrangement. This instrument is designed to measure both particle size and charge in a single cell. The particles are exposed to an electrical charge, causing them to move toward the appropriate electrode. By splitting and recombining the illuminating beam, an interference fringe can be generated inside the cell. As particles move through the fringes under the influence of the potential gradient, intensity fluctuations are produced whose frequencies are directly related to the velocity of the particles. These fluctuating intensities are detected by a photomultiplier and then processed by a spectrum analyzer in the

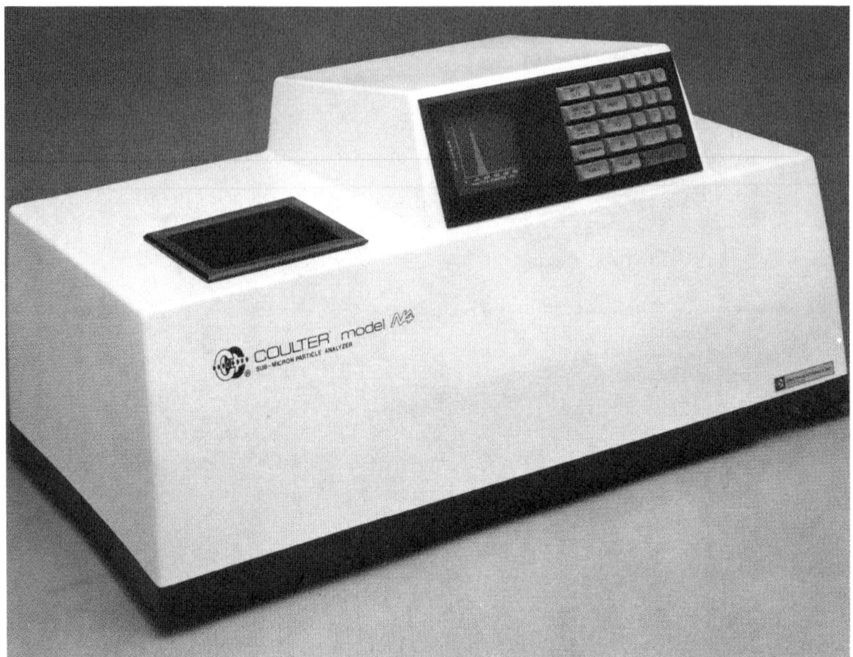

Figure 12. The Coulter Model N4 Sub-Micron Particle Analyzer. Reproduced with the permission of Coulter Electronics.

same manner as that used for PCS. This method of analysis is employed widely for the measurement of aerosol particles and is known as the Laser Doppler Velocimetry principle (51). The combination of two techniques to provide two fundamental measurements suggests that this instrument will enjoy wide applications in several different fields.

The fiber optic Doppler anemometer, FODA, has a similar computational method, but a radically different optical arrangement (Figure 15). This instrument pipes laser light down a fiber optic set into a suspension. As individual particles pass the end of the fiber, light is reflected back down the pipe and onto a photomultiplier, causing a Doppler shift. The major advantage of this device is that no cell is required for the analysis. The fiber top can be inserted into a relatively concentrated particulate system that is either static or flowing, and can be continually measured as a function of time. For this reason, this instrument may be applicable to the size analysis of systems undergoing dynamic changes with respect to their distribution. This instrument, described by Dyott et al. (52–53)

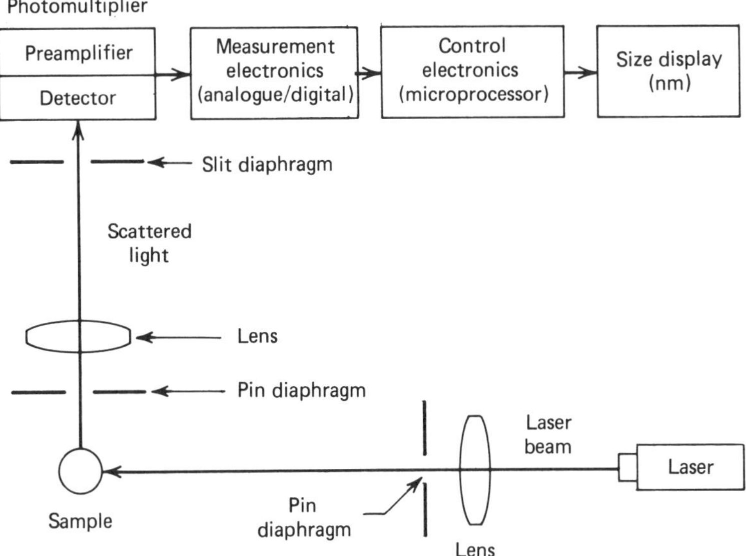

Figure 13.   Schematic of the Coulter Nanosizer. Reproduced with the permission of Coulter Electronics.

is currently available through the *Sira Institute*. Since the device is portable and robust, measurement is allowed in locations that are normally difficult to access, and it is likely that industrial applications for this instrument will be readily found. Single particles are effectively measured in relatively concentrated suspensions, often without the need for the prior dilution usually required for photon correlation spectroscopy, which is a unique feature.

9.   CENTRIFUGAL METHODS

Particle size distribution may also be determined by measuring the rate at which particles sediment through the system. Gravity sedimentation is not feasible for particles much below 5 μm (depending on the solid density), owing to the increased influence of Brownian movement, but this may be overcome by the application of a centrifugal field. Centrifugal size analytical methods have been used for many years to determine characteristics of submicron systems.

There are alternative methods of analysis. For example, in the two-layer method, a thin layer of a concentrated dispersion is added to the

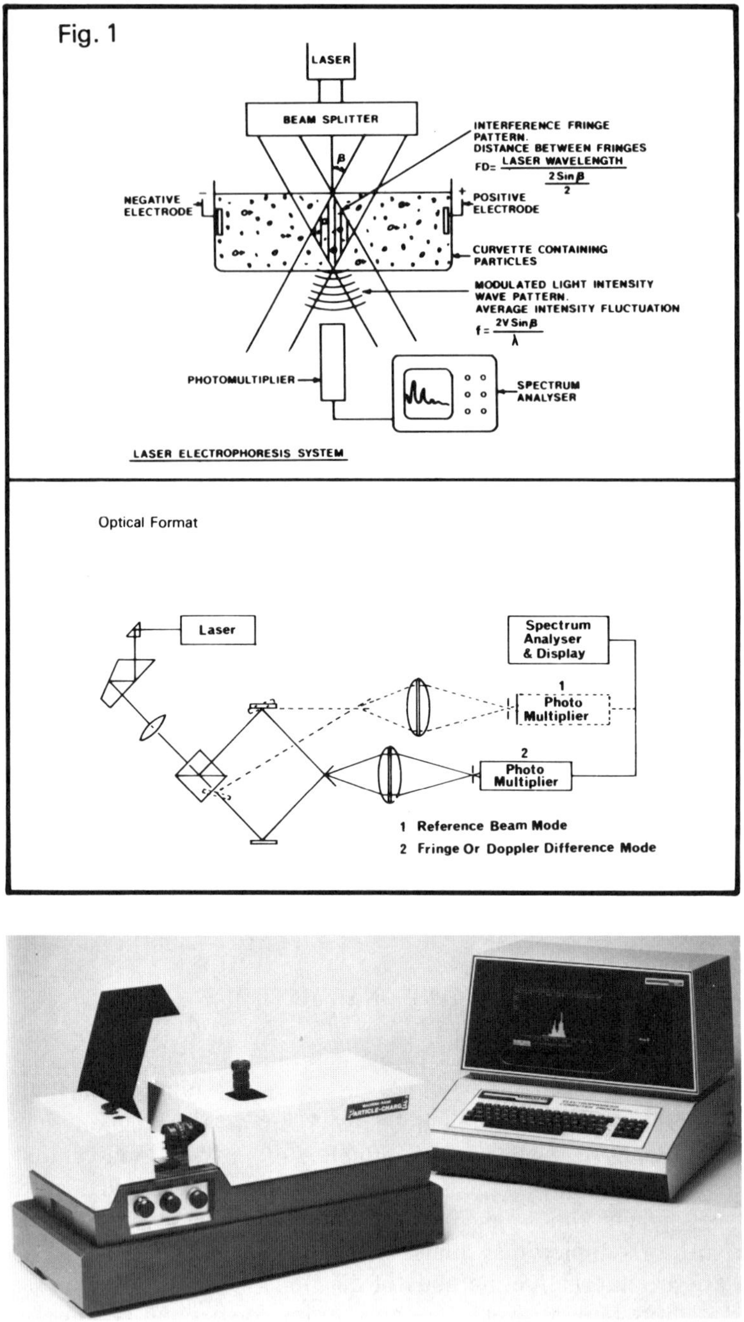

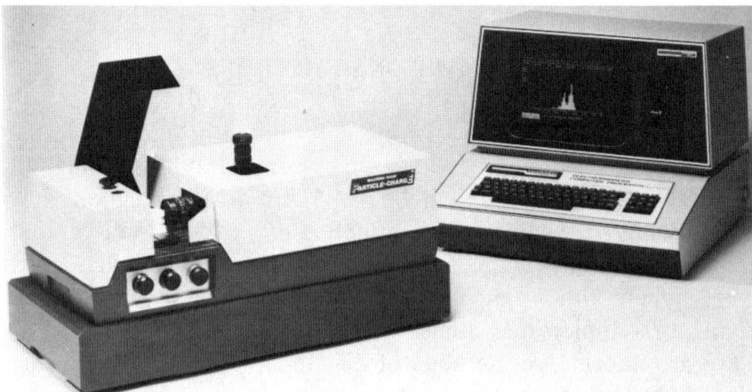

Figure 14. The Malvern Particle Charge Instrument. Diagram and photograph courtesy of Malvern Instruments Ltd.

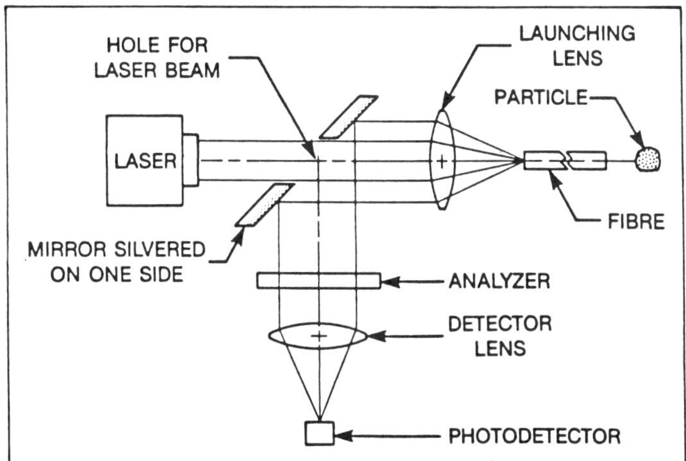

Figure 15. Diagram of the operating principle of the Fiber Optic Doppler Anemometer. Reproduced with the permission of Academic Press and R. B. Dyott, Andrew Inc.

top of a column of clear sedimenting liquid and is allowed to sediment through the column, the larger particles moving faster than the smaller. Alternatively, the suspension may be uniformly dispersed in the sedimenting liquid to form an initially uniform dispersion that is allowed to sediment as a function of time. The rate of change of concentration, at a point within the suspension, with time can then be related to the particle size distribution.

For convenience, methods of sedimentation analysis may be divided into two main categories: incremental methods and cumulative methods. In incremental methods, changes with time in the suspension concentration are measured at known depths in either fixed depth of fixed time modes of operation. In cumulative methods, the rate at which particulate material is settling out of suspension is measured. For the majority of centrifugal methods, the concentration of the sedimenting system is measured at a fixed point, called the analytical radius, as a function of time.

A single particle falling through an infinite fluid continuum will eventually fall at a terminal constant velocity determined by the size of the particle and the resistance to motion offered by the continuum. In a centrifugal field, this terminal velocity is not constant, because the strength of the centrifugal field is a function of the centrifugal radius. Measurement of the "terminal" velocity is the basis by which size may be deduced, provided the particles are moving in streamline motion. The relationship between drag and the movement of fluid around the falling particle may be reduced to the Stokes' equation. For fluid moving past a particle of diameter d_s, the ratio of the inertial transfer is described by the dimensionless parameter, the Reynolds number, R_E:

$$R_E = \frac{\rho_f u d_s}{\eta} \tag{17}$$

where ρ_f is the fluid density, η is the fluid viscosity, and u is the velocity. If $R_E \leq 0.2$, the fluid conditions are described as streamline or laminar, and the drag on the falling particle is due mainly to viscous force within the fluid. Particles with high densities or large particle diameters, for example, may be moving with velocities that exceed an R_E of 0.2, and in this situation are likely to enter the region of secondary or turbulent flow, when velocities are less predictable.

In a centrifugal field, Stokes' equation has the form:

$$u = \frac{(\rho_s - \rho_f)\omega^2 d^2}{18\eta}$$

$$= \frac{\ln(r/S)}{t} \tag{18}$$

where ω is the rotational velocity in radians/sec and t is the time (sec) required for a particle of diameter d_s to move from its starting point radius S to the analytical radius r. In the case of a particle whose density, ρ_s, is greater than that of the liquid, ρ_f, the starting radius, S, is assumed to be that of the inner liquid meniscus. However, in the case of many emulsion systems, if the density of the particle is less than that of the continuum, the particle will, in effect, float or move to the inner meniscus of the liquid. In this situation, the radius S is assumed to be the outer radius of the disk or, if tubes are used, the bottom of the tube.

Application of Stokes' equation makes certain assumptions that are not always achieved, and all of which are critical to the measurement of the size of the sedimenting particles. The first assumption is that the particle is spherical, smooth, and rigid. This is not always valid; thus, the diameter calculated is an equivalent of Stokes' diameter (d_{st}). It is also assumed that the particle terminal velocity is reached instantly, although calculation shows that a finite but small time is actually required before this condition is reached. The particle is also assumed to be moving without interference or interaction from other particles in the system. This assumption is only true in high dilutions to ensure considerable separation between the particles. Volume concentrations well below 1% is preferred.

Also, it is assumed that inertial effects are not present and that the fluid continuum only exhibits Newtonian flow characteristics. Since water is generally the dilution medium and particles are below 1 μm, these assumptions are usually valid.

The usual method of determining the concentration of particles in a suspension is by monitoring the attentuation a beam of radiation as a function of time. A major attraction of this detection system is that the sedimenting system is not disturbed during analysis. However, as noted with light scattering methods, the relationship between the particle concentration and attenuation of the beam becomes more complex and less certain as the particle diameter approaches the magnitude of the incident radiation wavelength. One solution has been to employ x-rays, but this approach is limited to materials that do not absorb the radiation. Generally, this is true for inorganic minerals and similar components, but has no application to materials such as polymers or hydrocarbon emulsions that are organic in nature.

The uses of white light sources and low speed centrifuges were reviewed by Groves and Freshwater (14). It was suggested that, if the sedimentary particles were large enough to interfere with the incident light, relatively unsophisticated equipment could be used, since the particles would separate according to Stokes' law. The authors also noted that although the quantitative expression for the particle concentration at a given time was not likely to be exact, and would result in an overemphasis of the finer particles in a given system, the relationship between time of sedimentation and attenuation would be characteristic of the system. In this manner, distributions could be compared, or fingerprinted.

If there are n particles per cm^3, each of projected area a, the fraction of light absorbed by the particles within a small thickness dl is given by:

$$-\frac{dI}{I} = Knadl \tag{19}$$

where I is the intensity of light entering the section (dl), dI is the quantity removed within the section, and K is the extinction coefficient. This equation integrates to:

$$\ln\left(\frac{I_0}{I_s}\right) = Knb \tag{20}$$

where I_0 and I_s are the intensities of the incident radiation and the radiation transmitted through the suspension respectively, and b is the path length of the cell. The extinction coefficient is a variable that is related to the particle size, and·may be determined theoretically or by calibration. Alternatively, K may be arbitrarily assigned a value of unity, although this distorts the shape of the distribution curve, especially below the dimensions of the wavelength of the incident radiation. Use of red laser light may, in part, compensate for this effect. As noted by Groves et al. (55), Rayleigh scattering by small particles is proportional to the inverse of the fourth power of the incident wavelength. A rigorous treatment of the relationship between the measured light transmittance and the particle concentration is not possible.

Theoretical treatment of data from a line-start centrifugal method is relatively straightforward, since it may be assumed that all particles start their sedimentation from the same radius. However, experimentally, line-start methods are notoriously difficult to run smoothly, since "streamers" or clouds of particles break through the original surface very readily. This "vortex" sedimentation, probably caused by local inequalities in density, invalidates the method, no matter how it is applied, and it is surprising to find papers recommending line-start techniques. This interfacial instability has been observed (56) at concentrations as low as 0.004% w/w.

Scarlett et al. (57) developed a three-layer technique in which a liquid of low density and viscosity was spread onto the inner liquid surface. The powder was then dispersed in a liquid of low surface tension and the suspension was injected into the annulus. A similar technique was patented by Jones (58, 59) and forms the basis of the method known as the "buffered line start" used in the Joyce–Loebl (Vickers Instruments, Inc.) centrifuge. Another similar technique was recently described by Allen and Khalili (60). Line-start methods are only suitable for particulate systems with densities in excess of that of the fluid continuum; however, the homogeneous suspension technique may be used irrespective of the density difference and, experimentally, is somewhat easier to utilize. The data manipulated is not so straightforward, because the suspension at time t contains a fraction of the particles in the radiation beam that are smaller than the largest particle d_{st}, having started from points between the known starting radius, S, and the analytical radius, r. These cannot be corrected for directly, although the iterative approximation method of Kamack (61, 62) gives a general solution readily calculated by standard computer methods.

Intuitively, it would appear that the particles obscure light with their cross-sectional areas so that the result of the analysis would be surface related. Treasure (63) demonstrated for the line-start method for a light beam of finite width that the obscuration was related to the mass of particles in the beam. More recently, Allan and Khalili (60) concluded that the homogeneous suspension method also provides a mass distribution

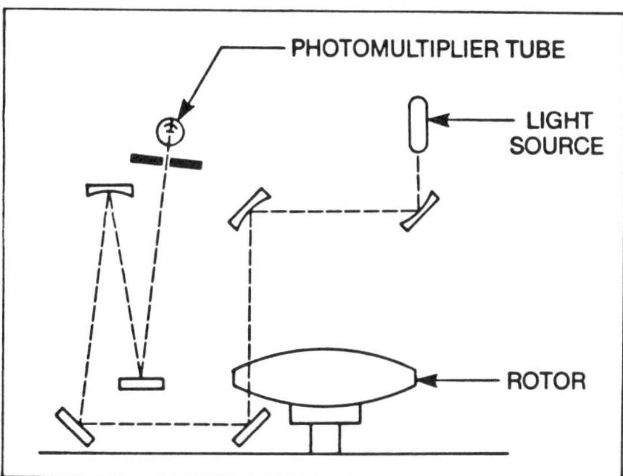

Figure 16. Schematic of the Beckman Ultracentrifuge. Reproduced with the permission of Beckman Instruments, Inc.

directly caused by a postulated linear proportionality between the extinction coefficient and the particle size. This conclusion must be established on a firmer experimental basis, especially for particles smaller than the incident beam wavelength.

In the past, the most frequently utilized submicron size analytical method was the analytical ultracentrifuge. This device has been applied to measuring physical properties of the larger molecules since its development by Svedberg in the 1920's. The commercial instruments have cells fitted in titanium or aluminum disks that rotate at speeds of up to 100,000 rpm in a partial vacuum to minimize heating effects caused by friction with air molecules. The techniques fall outside the scope of the present review since, in general, materials under investigation do not have definable surfaces. However, in the case of definable surfaced materials, Beckman offers a UV scanner attachment for their smaller laboratory models (Figure 16). This scanner tracks across the cell, producing a plot of the optical density (at 280 nm) as a function of the radial position along the cell. The scan takes approximately 1 min and the centrifuges are less than stable at speeds of 6000 rpm or less. For these reasons, this type of device may not be applicable for particles much larger than 500 nm, depending on the density difference between the solid and the liquid.

Low-speed disk centrifuges developed by Kaye and his associates (54, 64, 65) are more flexible in their application. These instruments have been utilized for the size analysis of materials between 50 nm and 5 μm, and generally used rotational speeds of up to 6000 rpm and white light optics.

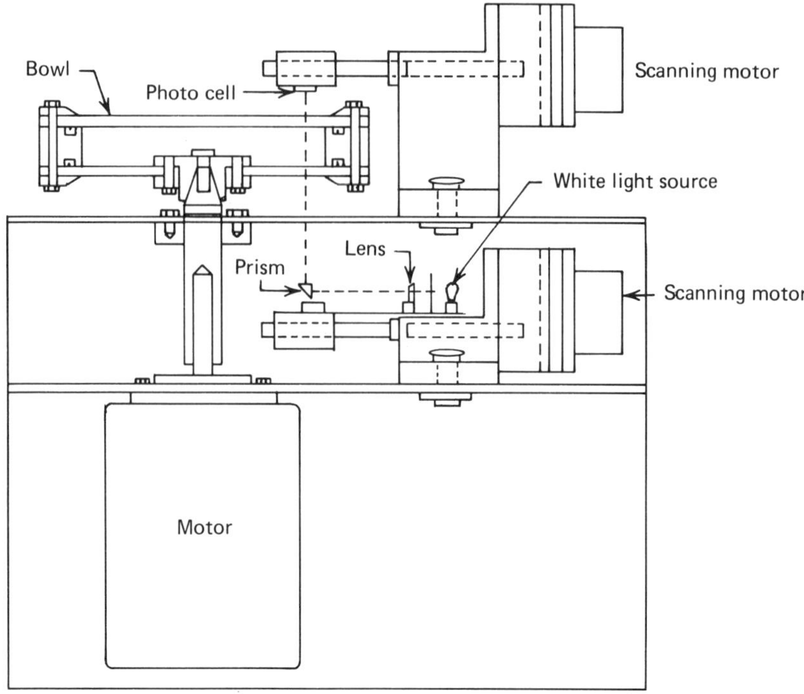

Figure 17.   A scanning photocentrifuge designed by Allen. From N. Stanley Wood and T. Allen (Eds.), ''Particle Size Analysis, 1981,' Chapman and Hall, London (1982), p. 288, used with permission.

A typical example is the Ladal photocentrifuge (Figure 17) (60), in which the light source and its associated detector is slowly tracked across the disk to provide a density profile during the sedimentation.

The Joyce–Loebl centrifuge (Figure 18), originally a sampling centrifuge, also incorporates a relatively simple white light source and detector to allow this instrument to act as a centrifugal analogue of a gravity photosedimentometer.

An alternative approach to the use of a white light source is provided by the Technord laser centrifugal photosedimentometer (Figures 19 and 20). This instrument uses laser light to determine the optical density at a point in the hollow rotating disk and would appear to be appreciably more sensitive than similar white light instruments. An early investigation using this instrument (66) showed that size analyses at different disk radii resulted in different size distributions from the same particulate system. This was attributed to insufficient centrifugal field being employed to

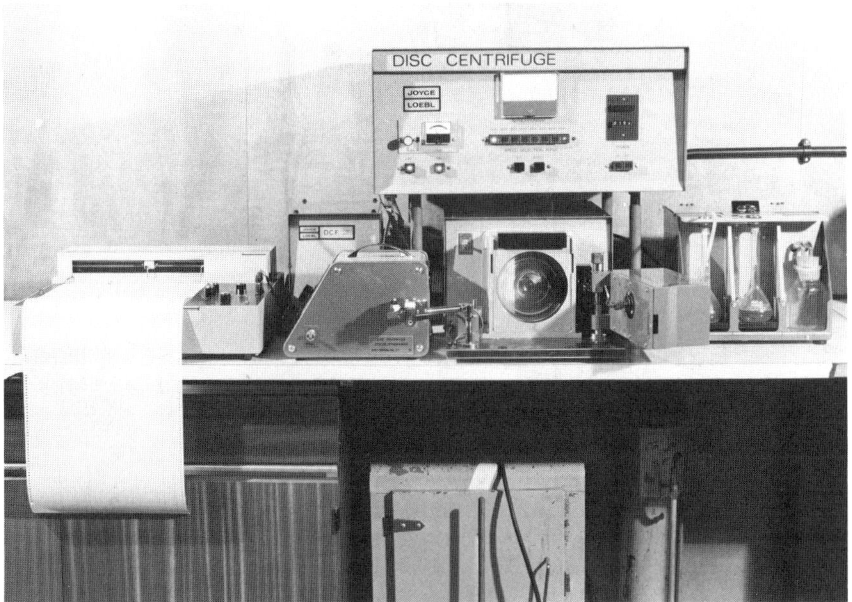

Figure 18. Joyce–Loebl Centrifuge. Reproduced with the permission of Vickers Instruments Inc.

overcome Brownian movement of the particles, especially the smaller particles at the lower end of this distribution. Thus, the distribution was distorted, since there could be no certainty that all of the particles in a system were actually moving in accordance with Stokes' Law predictions. To overcome this problem, the instrument has been modified to provide two simultaneous analyses on the sedimenting system at different radii (Figure 20). The two distributions are then matched and the data are adjusted to provide a distribution that would be obtained at the maximum centrifugal field used in the experiment, that is, at the edge of the disk. This approach was successfully applied to an emulsion system after correction for the hydrated layer around the triglyceride droplets, as shown in Table 1. A future version of this instrument will feature an electronically controlled method of determining the optical density profile across the disk. This must be obtained rapidly since, in any sedimenting system, the optical density changes with time. Mechanical scanning devices such as the Beckman Inc. and Ladal Ltd. instruments are of less value, since a relatively long time is required to scan the rotating disk.

In some situations, sampling centrifuges have been successfully employed where the concentration of particles must be determined by chem-

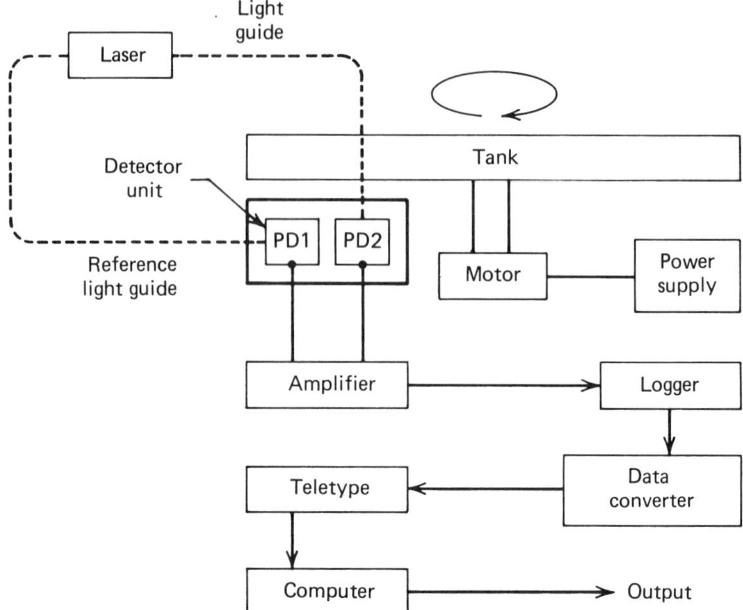

Figure 19. Layout of the Technord Laser Centrifuge Photosedimentometer. Reproduced with the permission of Technord Associates, Inc.

ical analyses. The suspension is allowed to run in the centrifuge disk for a known time and the supernatant is withdrawn for analysis. The best known sampling centrifuge is the Joyce–Loebl instrument, which features a hollow needle inserted isokinetically into the sedimenting suspension. Application of a vacuum enables a sample to be withdrawn at a fixed depth below the surface. Since insoluble colors and pigments are generally readily analyzed, the Joyce–Loebl instrument has been widely and successfully employed in the pigment industry. However, this instrument has been criticized (14), because of the inevitable suspension disturbance that must occur when the sampling probe tracks through the suspension. The Simon Carves (SimCar) centrifuge avoids this difficulty by having hollow tubes at a fixed radius inserted into the disk (Figure 21). This makes the instrument the centrifugal analogue of the Andreason pipette. In this latter device, samples are withdrawn from a gravity sedimentation system at a fixed or variable depth, using a pipette at appropriate time intervals (1).

Similar instruments are now available from Fritsch GmbH (Figure 22) and Ladal Ltd. These two devices have an advantage over the SimCar instrument in that smaller quantities of sample are required, because of

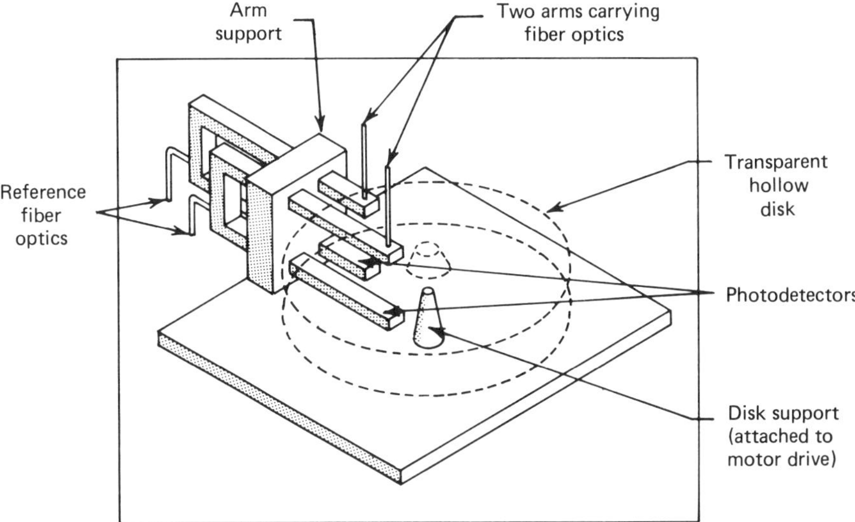

Figure 20.   Schematic of the Technord Laser Centrifuge Photosedimentometer. Reproduced with the permission of Technord Associates, Inc.

the disk's smaller volumes. The size distributions are built up by repeating the analyses and taking samples at different times. For this reason, a complete size distribution may take several hours, and successful application is critically dependent upon having a sensitive analytical method for the sedimenting material. These instruments are robust, relatively inexpensive, and unsophisticated. The main advantage that centrifugal photosedimentometers enjoy over sampling centrifuges is that the results are immediately available.

10.   ELECTRICAL ZONE SENSING DEVICES

The instruments based on the Coulter principle are capable of detecting and measuring submicron particles when operated under optimum conditions. When a particle passes through an orifice submerged in an electrolytic solution with two electrodes on either side, the resistance between the two electrodes is momentarily increased. The principle can be used to count particles as they pass through the orifice. The change in resistance caused by individual particles passing through the orifice is a measure of the electrolyte displaced by the particles and, hence, of the size. With appropriate electronic gating, the Coulter principle can be used to

Table 1.   Particle Size Distribution of Intralipid 10% using a Technord Laser Centrifugal Photosedimentometer

(A)   Size distributions obtained on samples dispersed in different concentrations of sodium chloride, extrapolated to conditions of maximum centrifugal field (at the rotational speed used).

Dispersant (M NaCl)	Mean Stokes' Diameter (d_{st} (50) (μm)	Slope (Logarithmic Probit Distribution) (σ)
0.35	0.175	1.13
0.43	0.225	1.22
0.51	0.285	1.14
0.59	0.320	1.19

(B)   Size distributions shown in (A) after correction for the presence of a hydrated layer of thickness $d(r)$ with a density of 1.0.

Dispersant (M NaCl)	$d(r)$ (nm)	d_{st} (50) (μm)	Slope (Logarithmic Probit Distribution)
0.35	150	0.45	1.14
0.43	125	0.45	1.29
0.51	80	0.45	1.16
0.59	75	0.46	1.21
Surface volume diameter by electron microscopy:		0.46	1.42

[a] Adapted from H. S. Yalabik and M. J. Groves, *Proc. Int. Cong. Pharm. Technol. (Paris)*, **2,** 28 (1977).

provide information on particle size that is uniquely dependent on volume and, within fairly broad limits, is not affected by the shape or nature of the particle.

Coulter Electronics, Inc. has a wide range of instruments based on this principle, as do other companies such as Particle Data, TOA, and Celloscope. From a historical point of view, the reviews by Rabinovich (67, 68) provide information on a variety of instruments and applications using the Coulter principle. Developments and improvements continue to be made in the electronics and the related glassware, but the basic operation

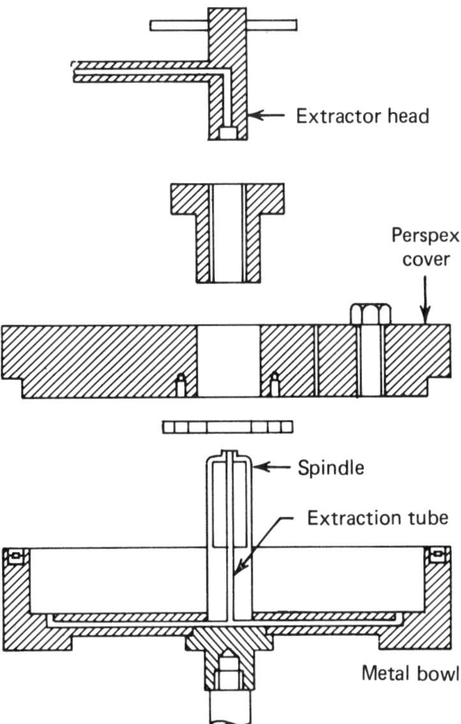

Figure 21. The Simcar Centrifuge. From T. Allen, *Particle Size Measurement*, by permission of Chapman and Hall, London.

remains much as it was in the early 1950's. Nonaqueous systems may also be sized, if a suitable conductive electrolyte medium is present.

There have been relatively few fundamental investigations of the underlying physical principles involved. Although it is clear that the instruments gauge a particle volume, it is less certain if this is based on the volume of electrolyte directly displaced or on that displaced by the particle sweeping through the orifice. In addition, the influence exerted by the particle surface is uncertain, especially if the particle itself is conducting, as might be the case for metal or metal oxide powders. Harfield (69) recently reexamined this problem and concluded that the dispersing medium also plays a part, because the medium could change the surface barrier potential and the emf applied to constant current instruments. This would result in negative and positive pulses as a conducting particle passes through the orifice. Experimentally, the effect may be reduced by increasing the concentration of the conducting electrolyte or by coating the

Figure 22. The Fritsch Analysette 21 pipette centrifuge. Photograph courtesy of Tekmar Inc. and the manufacturers, Fritsch GmbH.

surface with material of opposite electrical charge, such as a quaternary ammonium compound in the case of metal particles.

No fundamental error in the principle has been detected, although calibration of the device remains a problem (69–72) in some situations, especially if the particles under examination depart significantly from sphericity. Since orifices are available from Coulter with nominal sizes of 10-μm diameter, and the instruments can detect down to about 2% of the diameter, the theoretical lower limit of detection is around 200 nm. However, earlier claims (73) of detecting particles down to this size may be regarded with some caution, since the results may have been due to instrument idiosyncrasy (74). The instrument's lowest size limits are determined by thermal and electrical noise and by the ability of the electronics to separate the required signal pulse from the noise level. Realistically, it is generally accepted that the instrument may be used for

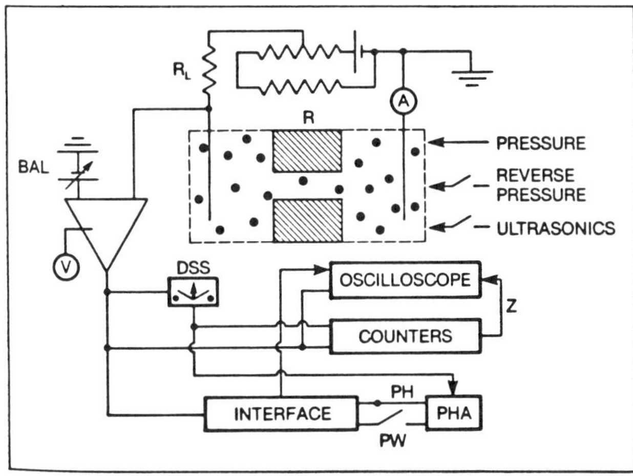

Figure 23. Schematic of the Nanopar analyzer: R_L = load resistance, A = nanoammeter, *BAL = balance potentiometer*, R = resistance of pore in nucleopore membrane, V = voltmeter, *PHA* = pulse height analyzer, *PW* = pulse width, and *PH* = pulse height. Reproduced with the permission of Academic Press and the authors (75).

particles down to 1 μm and will detect particles down to, and below, 500 nm. This does not necessarily mean that particles in the submicron region are accurately measured and characterized.

A significant loss of count is caused by the shadowing of small particles by larger ones when both are present at once (1). One way around this problem is to convey the particles through an orifice in a sheath of carrier liquid, enabling them to pass through one at a time. However, this is limited at present to > 1 μm particles. A modern instrument fitted with a 10-μm orifice and operated in a "quiet" electrical environment may be employed for the examination of submicron systems, although the electrolyte employed must be more highly concentrated and ultraclean. This condition is only obtained by continuous recirculation of solution through 0.1 or 0.2 μm pore-size membrane filters until the background count is at acceptable levels. Techniques used require special care to avoid extraneous contamination. The practical difficulty of making smaller holes in the sapphire wafer now used will obviously limit the reduction of this lower particle-size threshold.

The new Nanopar analyzer (75–77) addresses this issue. This device, shown in Figure 23, utilizes the Coulter principle, but has an orifice made from a Nuclepore filtration membrane. This membrane is prepared by passing cast polycarbonate film across a neutron source. Tracks left in the film are etched chemically to provide a precise pore density and pore size. In the Nanopar device, a single pore is exposed to act as an orifice with a vastly enhanced sensitivity to volume. Particles are made to pass through the pore under pressure, and ultrasonic irradiation is used to suspend the particles and free the pore from blockage. However, since blockage of such small pores is very easy, the instrument allows ready replacement of the pore. This factor probably renders the instrument more suitable for research and, at the time of this writing, the Nanopar is not available commercially. Practically, the Nanopar has been used to size viruses and bacteriophage particles down to 60 nm, and this limit may be lowered to 20 or even 10 nm. From an experimental point of view, the instrument also has the advantage of measuring the electrophoretic mobility of each particle as it passes through the orifice. For example, with this instrument, two lots of T-even bacteriophage were found to contain components that differed both in size and in electrophoretic mobility (78), information that otherwise would have been quite difficult to obtain.

11. MISCELLANEOUS METHODS

Undoubtedly, there are probably other methods for characterizing submicron particulate dispersion and emulsion systems that are not included in this survey. However, mention should be made of two methods that have application under some specialized circumstances.

The first is filtration, which initially would not be very promising. Filtration through stacks of Nuclepore membranes was used (70) to characterize liposomes prepared from aqueous dispersions of phosphatides. It was also used to characterize the particulate nature of xanthan gum (79) and to measure aggregates in solutions of guar gum (80). The method is unlikely to be very accurate or precise, but provides information that may be useful in the broad sense of characterization. One major problem is the occurrence of concentration polarization on the filter.

The other method involves measuring the surface area of a powdered solid. Surface area measurements provide little information about the spread of size around a mean, but are very sensitive to the presence of small particles in a system. Measurements may be taken by adsorbing an inert gas such as nitrogen, krypton, or argon onto the surface, or by passing a gas through a packed bed of the powder. The latter method tends to be limited to the larger (> 1 μm) particulate systems, but adsorption methods are widely applicable. In liquid systems, adsorption of dyes or fatty acids may be employed, since these can be readily measured.

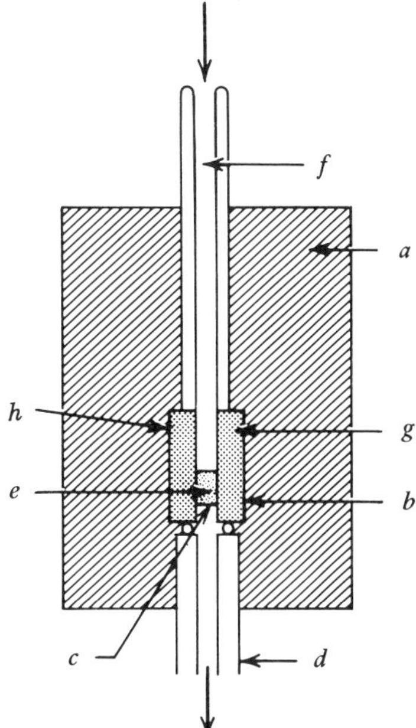

Figure 24. The Microscal flow microcalorimeter: a = metal block, b = calorimeter cell, c = gauze, d = outlet tube, e = adsorbent, f = central channel, g = thermistors, and h = reference thermistors. By permission of Microscal Ltd.

Flow microcalorimetry is a useful method for surface area measurement of some metal oxides and catalysts. Heat given out or taken up as an adsorbant is eluted from a solid surface is measured using thermistors in a flow-through cell, as shown in Figure 24. This device may also be applied to gases passing through the powder bed.

BET gas adsorption methods are widely employed for granular porous materials such as carbon blacks, charcoals, silica, and catalysts. All adsorption methods provide a measure of the total surface of a porous solid, both internal and external. Results are critically influenced by the depth to which an adsorbate molecule is capable of penetrating into the pores of the solid. It would, therefore, be anticipated that a nitrogen adsorption may provide a different result from, say, an experiment on the same material using krypton. Interpretation of the data depends on a knowledge of the area occupied on the solid surface by an adsorbate molecule and assumes that a monomolecular layer of adsorbate is established at the surface. Surface characterization methods are somewhat outside the scope of the present chapter and the interested reader is referred to such standard texts as those by Allen (1) or Gregg and Sing (81).

12. CONCLUSIONS

In recent years, the science and technology associated with the detection and measurement of submicron particulate systems have made considerable advances. Particles below the wavelength of visible light can now be sized and classified one at a time, in sufficient numbers to make a size analysis more meaningful within a reasonable time frame. The technologies associated with the production, handling, and application of submicron dispersions will, therefore, see some interesting developments in the not too distant future.

REFERENCES

1. T. Allen, *Particle Size Measurement*, 3rd ed., Chapman and Hall, London, 1981.

2. M. J. Groves, *Pharm. Technol.*, **4**(5), 80–94 (1980).

3. Clyde Orr, Jr., in M. J. Groves (Ed.), *Particle Size Analysis*, Heyden, London, 1978, pp. 77–100.

4. M. J. Groves, in J. W. Gorrod, (Ed.), *Drug Toxicity*, Taylor and Francis, London, 1979, pp. 101–122.

5. T. Fujita, T. Samaya, and K. Yokoyama, *Eur. Surg. Res.*, **3**, 436–453 (1971).

6. S. S. Davis, *Pharm. Technol.*, **5**(5), 70–88 (1981).

7. H. M. Patel and B. E. Ryman, in L. G. Knight (Ed.), *Liposomes: From Physical Structure to Therapeutic Applications*, Elsevier/North Holland, New York, 1981, pp. 409–436.

8. W. Carr, in M. J. Groves, (Ed.), *Particle Size Analysis*, Heyden, London, 1978, pp. 1–17.

9. F. B. Rosevear, *J. Am. Oil Chem. Soc.*, **31**, 628–639 (1954).

10. G. W. Gray and P. A. Winsor, *Liquid Crystals and Plastic Crystals*, Vols. I and II, John Wiley, New York, 1974.

11. M. J. Groves, *Chem. Ind.*, **17**, 417–423 (1978).

12. P. Becher, *Emulsion Theory and Practice*, 2nd ed., Chapman and Hall, New York, 1965.

13. International Union of Pure and Applied Chemistry, 1972.

14. M. J. Groves and D. C. Freshwater, *J. Pharm. Sci.*, **57**, 436–453 (1968).

15. T. A. Iranloye, Ph.D. Thesis, University of London (1981).

16. H. Heywood, in M. J. Groves and J. L. Wyatt-Sargent (Eds.), *Particle Size Analysis*, Society of Analytical Chemistry, London, 1972, pp. 1–18.

17. H. Heywood, *J. Pharm. Pharmacol.*, **15**, 56T–74T (1963).

18. G. Herdan, *Small Particle Statistics*, 2nd ed., Butterworths, London, 1960.

19. H. Small, *J. Colloid Interface Sci.*, **48**, 147–161 (1974).

20. H. Small, F. L. Saunders, and J. Solc, *Adv. Colloid Interface Sci.*, **6**, 237–266 (1976).

21. C. A. Daniels, S. A. McDonald, and J. A. Davidson, in P. Becher and M. N. Yudenfreund (Eds.), *Emulsions, Latices and Dispersions*, Dekker, New York, 1978, pp. 175–197.

22. C. A. Silebi and A. J. McHugh, *J. Appl. Polym. Sci.*, **23**, 1699–1721 (1979).

23. H. C. Van der Hulst, *Light Scattering by Small Particles*, Chapman and Hall, London, 1957.

24. J. W. Strutt, *Phil. Mag.*, **41**, 107–125 (1871).

25. J. W. Strutt, *Phil. Mag.*, **41**, 447–482 (1871).

26. Lord Rayleigh, *Phil. Mag.*, **44**, 28–53 (1897).

27. M. van der Waarden, *J. Colloid Sci.*, **9**, 215–230 (1954).

28. G. Mie, *Ann. Phys.*, **25**, 377–385 (1908).

29. F. Robillard and A. J. Patitsas, *Can. J. Phys.*, **51**, 2395–2402 (1973).

30. F. Robillard and A. J. Patitsas, *Can. J. Phys.*, **52**, 1571–1576 (1974).

31. F. Robillard and A. J. Patitsas, *Powder Technol.*, **9**, 247–255 (1974).

32. F. Robillard, A. J. Patitsas, and B. Kaye, *Powder Technol.*, **10**, 307–315 (1974).

33. *Atlas of Light Scattering Curves*, Wyatt Technology Co., Santa Barbara, California, 93130–3003.

34. P. J. Wyatt and D. T. Phillips, *J. Colloid Interface Sci.*, **39**, 125 (1972); J. N. Tand and H. R. Munkelwitz, *J. Colloid Interface Sci.*, **63**, 297 (1977); P. J. Wyatt, *Appl. Opt.*, **19**, 975 (1980).

35. E. J. Meeham and W. Beattie, *J. Phys. Chem.*, **64**, 1006–1021 (1960).

36. K. C. Yang and R. Hogg, *Anal. Chem.*, **51**, 758–763 (1979).

37. M. Kerker, *The Scattering of Light and Other Electromagnetic Radiation*, Academic, New York, 1969.

38. G. W. J. Lee and M. J. Groves, *Powder Technol.*, **28**, 49–54 (1981).

39. A. Lieberman, Proceedings of the European Symposium on Particle Characterization, Nuremberg, 1979, pps. 177–190.

40. J. H. Talbot in, M. J. Groves and J. C. Wyatt-Sargent (Eds.), *Particle Size Analysis 1970*, Society of Analytical Chemistry, London, 1972, pp. 96–100.

41. J. H. Talbot and D. J. Jacobs, in M. J. Groves and J. C. Wyatt-Sargent (Eds.), *Particle Size Analysis 1970*, Society of Analytical Chemistry, London, 1972, pp. 96–100.

42. E. C. Muly and H. N. Frock, Proceedings of the 1981 Particle Size Analysis Conference, Loughborough, in press.

43. P. N. Pusey, D. W. Schaefer, D. E. Koppel, R. D. Camerini-Otero, and R. M. Franklin, *J. Phys.*, **33**, C1 (1972).

44. J. C. Brown and P. N. Pusey, *J. Phys. D*, **7**, L31 (1974).

45. J. C. Brown, P. N. Pusey, and R. Dietaz, *J. Chem. Phys.*, **62**, 1136–1142 (1975).
46. S. Lacharojana and D. Caroline, *NATO Adv. Study Inst. Ser. B 1976*, **B23**, 499–514 (1977).
47. Ch. Gahwiller, *Powder Technol.*, **25**, 11–13 (1980).
48. C. A. Daniels and A. A. Etter, *Powder Technol.*, **34**, 113–119 (1983).
49. R. W. Lines and B. V. Miller, *Powder Technol.*, **24**, 91–96 (1979).
50. D. F. Darling and A. T. Shaw, Proceedings of the 1981 Particle Size Analysis Conference, Loughborough, in press.
51. P. Menon, *Am. Lab.*, **14**(2), 122–140 (1982).
52. R. B. Dyott, *I.E.E. J. Microwaves Opt. Acoust.*, **2**, 13–20 (1978).
53. D. A. Ross, H. S. Dhadwal, and R. B. Dyott, *J. Colloid Interface Sci.*, **64**, 533–542 (1978).
54. M. J. Groves, B. H. Kaye, and B. Scarlett, *Br. Chem. Eng.*, **9**, 742–746 (1964).
55. M. J. Groves, H. S. Yalabik, and J. A. Tempel, *Powder Technol.*, **11**, 245–255 (1975).
56. J. Beresford, *J. Oil and Col. Chem. Assoc.*, **50**, 594–603 (1967).
57. B. Scarlett, M. Rippon, and P. J. Lloyd, *Particle Size Analysis*, Society for Analytical Chemistry, London, 1967, pp. 242–261.
58. M. H. Jones, U. S. Patent No. 3,475,968 (1969).
59. M. H. Jones, *Proc. Soc. Anal. Chem.*, **3**, 116 (1966).
60. T. Allen and M. A. Khalili, Proceedings of the 1981 Particle Size Analysis Conference, Loughborough, in press.
61. H. J. Kamack, *Anal. Chem.*, **23**, 844–850 (1951).
62. H. J. Kamack, *J. Phys. D Appl. Phys.*, **5**, 1962–1968 (1972).
63. C. R. G. Treasure, Technical Paper No. 50, Welwyn Hall Research Association, 1964.
64. B. H. Kaye and M. R. Jackson, Paper Presented at 151st A. C. S. Meeting, Pittsburgh, March 1966.
65. B. H. Kaye, British Patent No. 895,22,222 (1962).
66. M. J. Groves and H. S. Yalabik, *Powder Technol.*, **12**, 233–238 (1975).
67. F. M. Rabinovich, *The Application of Conductimetric Particle Counters to Medicine*, Medical Publishing House, Moscow, 1972.
68. F. M. Rabinovich, *Conductimetric Methods of Dispersion Analysis*, Leningrad, 1970.
69. J. G. Harfield, Proceedings of the 1981 Particle Size Analysis Conference, Loughborough, in press.
70. J. G. Harfield and P. Knight, Proceedings of the 1981 Particle Size Analysis Conference, Loughborough, in press.
71. P. J. Lloyd, Proceedings of the 1981 Particle Size Analysis Conference, Loughborough, in press.

72. C. M. L. Atkinson and R. Wilson, Proceedings of the 1981 Particle Size Analysis Conference, Loughborough, in press.

73. H. E. Kubitschek, *Nature (London)*, **182,** 234–235 (1958).

74. R. W. Lines in *Particle Size Analysis*, Society of Analytical Chemistry, London, 1966, p. 134.

75. R. W. DeBlois, C. P. Bean, and R. K. A. Wesley, *J. Colloid Interface Sci.,* **61,** 323–335 (1977).

76. R. W. DeBlois, E. E. Uzgiris, D. H. Cluxton, and H. M. Mazzone, *Anal. Biochem.,* **90,** 273–288 (1978).

77. B. I. Feuer, E. E. Uzgiris, R. W. DeBlois, D. H. Cluxton, and J. Lenard, *Virology,* **90,** 156–161 (1978).

78. F. C. Szoka and D. Papahadjopoulos, *Proc. Natl. Acad. Sci., USA,* **75,** 4194–4198 (1978).

79. G. Holzworth and E. B. Prestridge, *Science,* **197,** 757–759 (1977).

80. P. J. Whitcomb, J. Gutowski, and W. W. Howland, *J. Appl. Polym. Sci.,* **25,** 2815 (1980).

81. S. J. Gregg and K. S. W. Sing, *Adsorption, Surface Area and Porosity*, Academic Press, N.Y., 1967.

CHAPTER

3

PARTICLE SIZING USING PHOTON CORRELATION SPECTROSCOPY

BRUCE B. WEINER

Brookhaven Instruments Corporation
Ronkonkoma, New York

1. INTRODUCTION

The large number of techniques available for particle sizing is a testimony to the enormous number of particle sizing applications (1). However, no single technique covers all sizes of interest, nor, within a given size range, is any single technique capable of handling all sample preparations. If the size range is restricted to below a few microns, and to particles suspended in liquids, the number of techniques dwindles to a few: sedimentation/ centrifugation, electric and light-zone blockage, hydrodynamic chromatography, sedimentation flow field fractionation, and light scattering. The last technique encompasses many different variations, and it is the purpose of this chapter to review, with respect to particle sizing, a relatively new variation, photon correlation spectroscopy (PCS). Several books on the theory and practice of PCS and related techniques already exist (2– 7).

Experiments utilizing PCS were first performed in the 1970's (8), and they were extensions and refinements of the earlier work on light-beating spectroscopy done in the 1960's (9). Both types of experiments are concerned with the time dependence of the fluctuations in the scattered light intensity. It is the integrated or average intensity that is most commonly

93

referred to when scattering experiments are discussed and not the intensity fluctuations. To clarify the distinctions, and because various features of the intensity of scattered light are pertinent to the interpretation of data from polydisperse samples, it is worthwhile to review briefly the main features of the theory of the intensity of scattered light.

In the last century, experiments by Tyndall (10) on the scattering of white light from suspensions of very small particles in a gas were investigated theoretically by Rayleigh, using the then relatively new theory of electromagnetic radiation (11, 12). The oscillating electric field in a beam of light passing through a dielectric medium induces oscillating electric dipoles in the polarizable particles constituting the medium. These oscillating dipoles, which reradiate the light like the antenna of a radio transmitting station, are the source of the scattered light.

Rayleigh theory is correct for single particles with a size, d, very much smaller than the wavelength of light, λ. However, in condensed media, the theory breaks down because of the destructive interference between scattered fields from different particles.

Smoluchowski (13) and Einstein (14) reformulated the Rayleigh theory from one concerned with polarizability and the resultant induced dipole moment per particle to one concerned with polarizability and dipole moment per unit volume of the dielectric medium. Since, at any finite temperature, there are thermal fluctuations in all the properties of any medium, the polarizability per unit volume at any position and time can be thought of as the sum of a constant part and a fluctuating part. The constant part gives rise to the familiar optical effect of refraction, whereas the fluctuating part gives rise to scattering.

The intensity of the scattered light, according to fluctuation theory, may be calculated from the mean-square fluctuations in the density and entropy for a pure fluid or the concentration for a suspension or solution. Complete descriptions of this highly successful theory, and its extensions to larger particles, are available elsewhere (15, 16). Of interest here is the result for the time-averaged intensity scattered from dilute suspensions or solutions of particles with sizes smaller than λ. (The scattering from the pure suspending medium is much smaller than that from the suspended particles.) With reference to Figure 1, the result is:

$$\langle I_s(q) \rangle = KNM^2 P(\theta)B(c) \tag{1}$$

where $\langle I_s(q) \rangle$ is the time-averaged scattered intensity from the particles; q is the wave vector amplitude of the scattering fluctuation; K is an optical constant; N is the number of particles contributing to the scattering; M

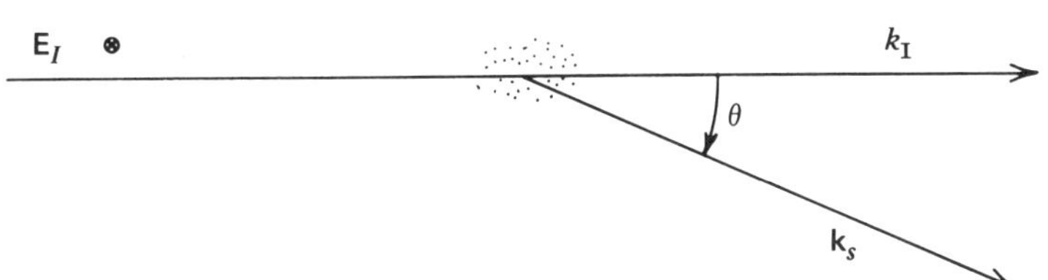

Figure 1.   Light with incident electric field **E**_I_ polarized perpendicular to plane defined by incident wave vector **k**_I_ and scattering wave vector **k**_s_ where **q** = **k**_s_ − **k**_I_.

is the mass of a particle; P is the particle form factor; θ is the scattering angle; B is the concentration factor; and c is the particle concentration.

The concentration fluctuations that give rise to the scattering are fluctuations in space and time. Prior to the availability of lasers and the technique of light-beating spectroscopy, only the spatial fluctuations were investigated. The time-averaged intensity arises from the spatial fluctuations.

Fluctuation theory has shifted the physically simpler notion of scattering from induced dipole oscillators to scattering from thermally excited fluctuations. These fluctuations may be decomposed into various frequencies and, at any angle, the scattering is due to a particular fluctuation represented by the wave vector q, with the amplitude given by

$$q = \frac{4\pi n}{\lambda_0} \sin \frac{\theta}{2} \tag{2}$$

where n is the index of refraction of the suspending medium.

Equation 2 is the Bragg condition; however, here the scattering is not from a lattice with a regular interatomic spacing, but from a thermally excited fluctuation with wavelength $\lambda_q = 2\pi/q$.

Since the mass of a spherical particle is proportional to its volume and, hence, its diameter cubed, the M^2 term in Equation 1 gives rise to a d^6 factor in the scattering. This has important consequences for interpreting PCS measurements on polydisperse samples.

The particle form factor $P(\theta)$ is a result of intraparticle interference for finite size particles. Formulas for P exist for simple shapes and sizes less than about $\lambda/2$ (16). Beyond this, and when the relative index of refraction of particle and medium differ significantly, the result for the scattered intensity must be calculated from the more complicated Lorenz–Mie theory. Fortunately, in the limit $\theta \rightarrow 0$, the particle form factors, even for large particles, go to unity. For this reason, angular PCS measurements with extrapolation to zero angle are required for polydisperse samples with large particles.

The concentration factor $B(c)$ is a result of interparticle effects. In the limit $c \rightarrow 0$, $B(c) \rightarrow 1$. Fortunately, photon detection is such a sensitive technique that measurements can often be made in very dilute suspensions where $B(c) = 1$. Otherwise, it is necessary to make measurements as a function of concentration with extrapolation to $c = 0$.

When the particles are macromolecules and the suspensions are true solutions, Equation 1 is extremely successful for determining molecular weight, radius of gyration (a size parameter in the form factor), and second virial coefficient (a parameter in the concentration factor) from the integrated or time-averaged intensity of scattered light. In the next section, the useful information on particle size contained in temporal fluctuations is discussed.

2. TIME-DEPENDENT LIGHT SCATTERING

In 1964, Pecora (17) showed that temporal fluctuations, which give rise to a broadening in the frequency of the scattered light, contain information on the motion of the particles. In particular, the translational diffusion coefficient D_T can be calculated from the linewidth Γ (defined as the half-width at half-maximum) of the scattered light. Since there are well-known relationships between D_T and particle size for simple shapes, particle sizing is possible.

The frequency of visible light is on the order of 10^{14} Hz; the linewidths of interest vary from 1 Hz to 1 MHz. Resolutions of 1 part in 10^9–10^{14} are not obtainable in the visible with standard techniques. Fortunately, the ideas of optical-mixing spectroscopy or light-beating spectroscopy were also pioneered (18) in the mid-1960's. The idea is to downshift the very high center frequency in the visible to zero. This is accomplished by optically mixing or beating coherently the scattered light with a tiny portion of the main beam (local oscillator) or with itself (self-beating). In both cases, a difference spectrum, centered at zero frequency, is generated, allowing the use of conventional spectrum analyzers. The idea of mixing or beating has its analogue in the modulation of ordinary radio frequency signals. However, the technique described here requires a light source of high coherence. Thus, the commercial availability of continuous wave lasers, at about the same time, was necessary for measurements.

Although the spectral broadening was initially measured with spectrum analyzers, correlators are, for the type of measurements described here, more practical, and they have been used almost exclusively since the early to mid-1970's. Since the Fourier transform of the power spectrum is equal to the autocorrelation function, the information obtained is the

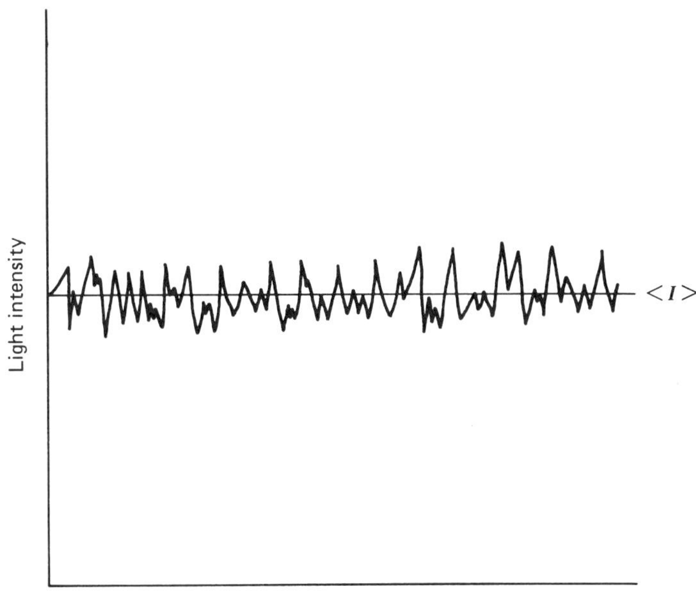

Figure 2. Scattered light intensity as a function of time (average plus fluctuations).

same. However, spectrum analyzers are analogue instruments with bandwidths that introduce errors into the measurements at low frequencies. However, a good digital correlator can measure very low light levels with equivalent bandwidths that are orders of magnitude smaller than a spectrum analyzer. In very high light levels (count rates above a few MHz), where photon correlators have difficulty resolving pulses, a real-time spectrum analyzer has some advantages. Also, in some laser doppler experiments on particles in uniform motion (electrophoresis) where a frequency shift rather than a broadening is desired, a spectrum analyzer would be more useful than an autocorrelator.

The theory of time-dependent or dynamic light scattering can be found in the book by Berne and Pecora (3). Here it is appropriate to review a few basic concepts.

Figure 2 shows schematically the intensity of scattered light versus time. It consists of a time-averaged part and a temporally fluctuating part. Since the particles are in constant, random or Brownian motion, their positions are continually changing. The scattered electric field, which is a function of particle position, is also constantly changing. Since the intensity is proportional to the square of the electric field, it too is fluctuating in time.

The dynamical information of interest (diffusion coefficient) is contained in the fluctuations, and fluctuations are conveniently described by time-dependent correlation functions. The autocorrelation function is defined as follows:

$$\langle I(0)I(\tau)\rangle = \lim_{T\to\infty} \frac{1}{T} \int_0^T I(t)I(t-\tau)\,dt \tag{3}$$

where the intensity I has, in general, different values at time t and $t - \tau$. T represents the total experiment duration, over which the product of the intensity with delayed versions of itself is averaged. The starting time of the experiment is unimportant for this type of stationary (in time) system and, therefore, t and $t - \tau$ can be replaced by 0 and τ.

Since particle motion is correlated over a time that is small compared to a characteristic fluctuation time (the $1/e$ point in the correlation function), the product of intensities is high at short times and the autocorrelation function is equal to $\langle I^2 \rangle$, the mean-square intensity at $\tau = 0$. At long times, all correlation is lost and

$$\lim_{\tau\to\infty} \langle I(0)I(\tau)\rangle = \langle I(0)\rangle\langle I(\tau)\rangle = \langle I\rangle^2 \tag{4}$$

That is, at long times the autocorrelation function has decayed and is equal to the square of the average intensity. The question, then, is how do the fluctuations decay?

In a classical diffusion experiment, two separate phases, one of pure solvent and one with a finite concentration of solute, are allowed, at $t = 0$, to come to equilibrium. This macroscopic relaxation back to equilibrium is described by Fick's first law of diffusion,

$$\frac{\partial c}{\partial t}(\mathbf{r}, t) = -D_T \nabla c(\mathbf{r}, t) \tag{5}$$

where $\mathbf{r}$ is a vector representing the spatial dependence of the concentration. In this way, the diffusion coefficient is introduced into the dynamical description of the system under study.

Even at equilibrium, concentration gradients exist as a result of the thermal fluctuations. The Onsager regression hypothesis states that these microscopic concentration fluctuations relax back to equilibrium according to the same equations as a macroscopic gradient. When these ideas are used to calculate the intensity fluctuations, the intensity autocorrelation function decays exponentially with time. The characteristic decay time depends on D_T. Therefore, D_T appears as the principle parameter of interest in the autocorrelation function.

Equation 3 defines the autocorrelation function of an analogue signal. In practice, the integral is approximated by a sum over discrete values. Furthermore, scattered light consists of pulses or photons. Intensity corresponds to the number of pulses per unit time. If $n(t)$ represents the number of pulses registered between $t - \tau$ and t, then the photon–autocorrelation function can be written as

$$C(\tau) = \langle n(0)n(\tau) \rangle = \lim_{N \to \infty} \frac{1}{N} \sum_{j=1}^{N} n_j n_{j-m}, \qquad m = 1, 2, 3, \ldots M' \quad (6)$$

where M' is the number of points at which the correlation function is measured, and the time axis has been divided into equal increments of time Δt. This quantity is called the sample or delay time and all other times are related to it. The averaging time T is now characterized by the total number of samples N, where $T = N\Delta t$. Also, $t = j\Delta t$ and $\tau = m\Delta t$.

Using the results of Jakeman and Pike (19) for the self-beat, photon autocorrelation function from a system of rigid, monodisperse, compact particles, one obtains

$$C(\tau) = \langle n \rangle^2 [1 + b \exp(-2\Gamma\tau)] \quad (7)$$

where

$$\Gamma = D_T q^2 \quad (8)$$

and b is an experimental constant. It is assumed in the derivation of Equation 7 that the average number of particles in the observed scattering volume is large. This is called the Gaussian approximation and is satisfied for all but the most dilute solutions or when the incident beam is highly focused.

Figure 3 is a plot of Equation 7. The baseline $\langle n \rangle^2$ is proportional to the square of the total intensity, and it can be determined experimentally. The constants b and Γ are determined by fitting the measured function to a single exponential; D_T is then calculated using Equations 2 and 8. If the shape of the scatterer is known or assumed, then a particle size can be determined. For instance, the Stokes–Einstein equation for a sphere is

$$D_T = \frac{kT}{3\pi\eta d} \quad (9)$$

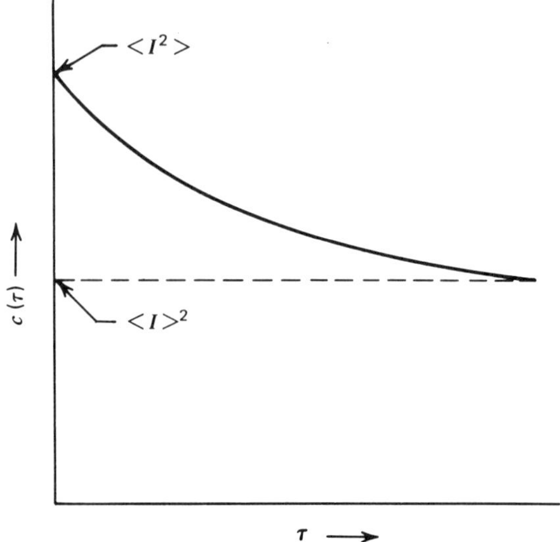

Figure 3. Autocorrelation function of scattered light intensity.

where k is the Boltzmann constant, T is the absolute temperature, and η is the viscosity of the medium in which the particles of diameter d are suspended.

In the following sections on instrumentation and operating conditions, the implications of Equations 7–9 are explored in terms of particle sizing. However, the diffusion coefficient obtained is independent of the shape assumption necessary for particle sizing.* PCS is now the preferred technique for measuring D_T in the submicron range.

3. INSTRUMENTATION

Figure 4 shows a typical instrument configuration: source optics including laser, goniometer with sample cell assembly, detection system, and combination signal processor/data analyzer. Each of these components is described in more detail below.

There are two varieties of lasers in general use for PCS measurements: helium–neon and argon–ion. Table 1 summarizes some pertinent features of each.

---

* For asymmetric, nonrigid particles, there are contributions to the autocorrelation function from rotational and intramolecular motions.

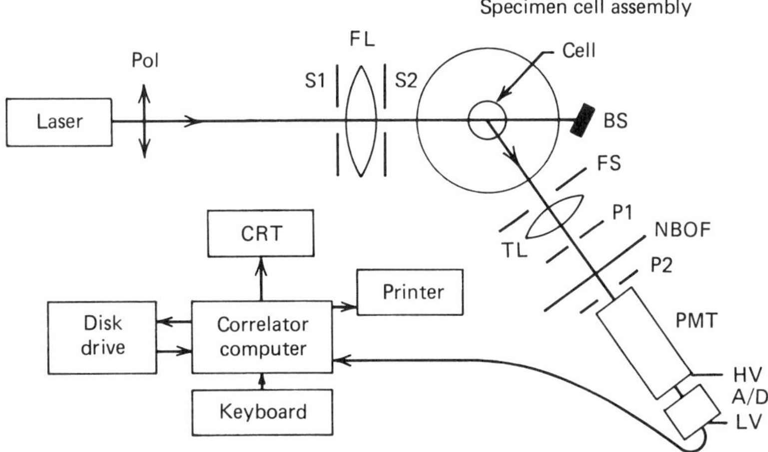

Figure 4. Typical multiangle instrument configuration. Pol indicates vertically polarized light; apertures S1 and S2 block stray light; FL is a focusing lens; BS is a beam stop; FS is a field stop for the transmitting lens, TL; pinholes P1 and P2 define the coherence and detected areas as well as the angular acceptance angle. A narrow-band optical filter (NBOF) allows only laser light to register on the photomultiplier tube (PMT), which requires a high voltage (HV) power supply. Pulses are amplified and discriminated (A/D) before entering the correlator. A low-voltage (LV) supply powers the A/D combination.

Small, air-cooled, Ar–ion lasers are available, but the power is only 5–10 mW and the cost is about $8K. Helium–cadmium lasers (442 nm, 5–40 mW, $4K–9.5K) are also sometimes used. Until recently, the price performance ratio for these lasers was too high to justify most measurements. For very weak scattering, which is not often the case in particle sizing applications (except in very dilute solutions of macromolecules),

Table 1. Comparison of Lasers for Use in Particle Sizing with PCS

	He–Ne	Ar–Ion
Wavelength	632.8 nm	488.0 and 514.5 nm
Power	5–15–50 mW	0.5–2 W
Stability	Excellent	Fair
Cost	$1K, $4.6K, $14K	$15K–$20K
Maintenance	Very little	Tube replacement
Size	Small, Medium, Large	Large
Cooling	Air	Water

an Ar–ion laser of high power is necessary; otherwise, a modestly powered He–Ne laser is sufficient.

Light polarized vertically to the scattering plane (Pol) defined by the incident and scattered wave vectors is necessary, and lasers that incorporate vertical polarization of at least 500:1 are sufficient.

A focusing lens (FL) is necessary for good quality measurements. The time-dependent part of the signal in a PCS measurement is proportional to the coherence area, which is increased by reducing the diameter of the nominal 1 mm laser beam to about 100 μm in the scattering volume. Apertures (S1, S2) help block stray light.

Sample cells are typically cylindrical with 8–10 mm path lengths. Sample volume is on the order of 1–3 cm^3. For the best angular measurements, selected glass or quartz cells of uniform wall thickness are annealed and highly polished. For routine work, lesser quality glass cells or even some plastic cells can be used. It is easier to make a more optically perfect square cell than a cylindrical cell; however, a Snell's law correction for refraction must be applied to obtain the true scattering angle, except at 90°. Also, some angles are prohibited by the corners of the cell and, sometimes, by total internal reflection at Brewster's angle. Inexpensive, square plastic cells can be used, but they are typically restricted to 90° scattering.

The greatest problems in light scattering are from flare light and dust. Flare light caused by surface imperfections is often much more intense than scattered light from the sample. Flare at air–cell interfaces is often the worst offender. As the scattering angle is reduced, the detector sees more and more light. By surrounding the small-diameter cell with an index-matching liquid contained in a large-diameter vat, smaller angles can be attained before flare from the entrance/exit windows of the cell or vat becomes a problem. Index matching allows the use of less high-quality cells, because imperfections on the outer cell walls are reduced. Flare from the entrance/exit windows of the vat is also a problem, and the optical quality and construction of the vat must be considered if good, angular measurements are required. Because index-matching liquids contain dust, it is necessary to filter the liquid. The effects of flare and dust (in the sample as well as in the index-matching liquid) on the measured autocorrelation function are discussed in Section 5.

Temperature measurement is necessary primarily to establish the correct viscosity, which is obtained from tabulated values. For room temperature measurements, during which the temperature of the sample varies by a few tenths of a degree or less, no temperature control is necessary. However, for temperature studies, or to reduce viscosity to an acceptable level, or for long-time measurements, it is necessary to control the cell temperature. Viscosities can change by several percent per degree. The relative error in the calculated diameter is proportional to the relative error in the viscosity. Care must be taken not to introduce temperature gradients, since this may cause flow. The theory assumes random, not uniform, motion of the particles.

A beam stop (BS) is useful for blocking the intense, transmitted beam. If suitably designed, it may also be used for measuring the transmitted beam in extinction experiments or for monitoring main beam intensity.

The overall angular range of a system like the one in Figure 4 is typically 10–160°. Below 20°, stray light, dust (in the sample as well as in the index matching liquid) and cell/vat imperfections become increasingly troublesome. For angular measurements, an automated system with a stepping motor is very convenient.

The detection optics consists of a field stop (FS), transmitting lens (TL), primary pinhole (P1), narrow-band optical filter (NBOF), and selectable pinholes (P2). Additionally, it is convenient to view the scattering volume and P1 for ease of alignment. The narrow band optical filter (NBOF) is included to reduce room light. The diameter of P1 is typically 100–200 μm, and P2 is chosen to match the coherence area of the scattered light reaching the photomultiplier tube. The selection of P2 is discussed in Section 4.

Properly selected photomultiplier tubes (PMT) are crucial for the best measurements on very small particles and at high angles where the sample times are small. If the PMT has inherent correlation owing to afterpulsing that is significant over the time range of interest, then the calculated size parameters may contain significant systematic errors. Afterpulsing is often caused by residual gas molecules in the tube which, when ionized, result in the detection of pulses trailing in time the main pulse.

Signals from the PMT are first amplified then discriminated to produce a uniform pulse train varying only in the time between pulses. All the information about the scatterers is contained in these pulses, which are fed into the correlator.

Figure 5 is a simplified diagram showing how a digital, real-time, multibit correlator works. During every sample time, the number of pulses applied to the input is counted by a 4-bit (up to 15 photopulses) counter. At the end of the period, the value in the counter is entered into the first stage of the shift register and the previous values in the shift register are shifted by one stage. After a time equal to $(j - m)\,\Delta t$, the mth stage contains the value n_{j-m}. Each stage corresponds to one correlator channel.

Also, during every sample time, the number of pulses n_j applied to the input are multiplied by the values in each of the shift register stages and

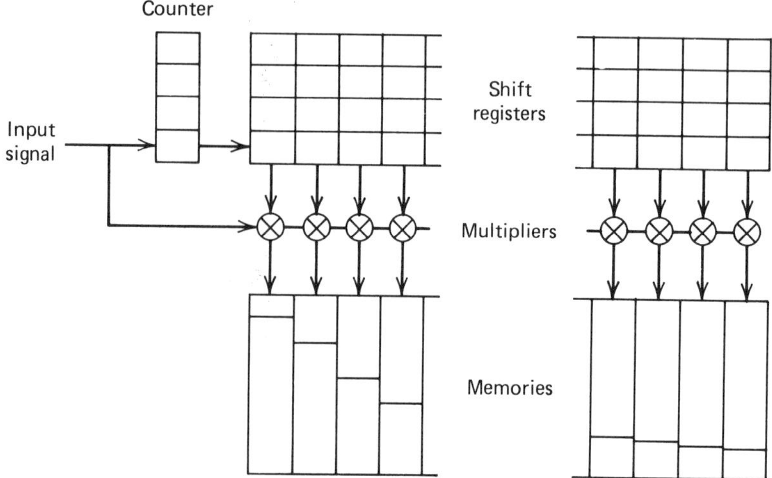

Figure 5.   Block diagram of digital autocorrelator.

added to the hardware memories. After a time equal to $N\Delta t$, the contents of the mth correlator channel contain the value

$$C(m\Delta t) = \sum_{j=1}^{N} n_j n_{j-m} \tag{10}$$

where n_j represents the number of times the contents n_{j-m} of the mth stage of the shift register is added into the mth hardware channel memory.

This method allows an $n \times 4$-bit approximation to the complete correlation function obtained by counting and multiplying every pulse during a sample time. The approximation becomes exact as the number of samples N grows large. In many cases of practical interest, the number of pulses/sample time is either less than 15 or can be reduced by prescaling electronically or by reducing the signal intensity before the PMT. Complete multiplication of every pulse at all sample times and all count rates is considerably more expensive at the present time.

Sometimes the channel memories consist of hardware counters with software memory to catch the overflows. Although less expensive to construct, this can limit the maximum count rate achievable without missing an overflow. The effect is noticeable, typically, above a 0.5 MHz count rate with 128 real-time correlator channels.

Analogue detection is not as sensitive as single photon detection. Analogue correlation introduces noise, which a digital correlator does not. A photon correlator that operates in either a batch mode or off-line increases significantly the total measurement time. Software multiplication limits seriously the minimum sample time available. For these reasons, using a hardwired, digital correlator operating in real-time is the most efficient way to process the signal (20).

To obtain size information, the data in the correlation function must be processed. Data may be transferred to an external computer, or processed internally. If an internal microprocessor is used, then either the analysis program is fixed in read-only memory or it is accessible via the programmable features of the internal computer. The latter method is more flexible; programs can be changed and updated much more easily as new analysis techniques become available. It includes the simplicity of the former method, however, because any single program can be run automatically.

In Section 4, the various criteria for setting experimental operating conditions is explored.

4. MEASUREMENT CONDITIONS

Besides laser power and wavelength and temperature control, several other measurement conditions must be established to optimize the measurement.

The total count rate (number of photons/sec) should be below 10 MHz. Above this level, combined dead-time effects yield an apparent count rate that is lower than the true rate. A more stringent requirement in the type of 4-bit machine outlined in the previous section is to keep $\bar{n}$, the average number of pulses/sample time, below 15.* Above this level, the correlator is clipping the signal, and the function produced is very nearly a straight line devoid of the exponential decay.

Count rates can be reduced by lowering the concentration, lowering the laser output power, changing the scattering angle, choosing a smaller detection aperture, using a neutral density filter, or electronically prescaling the correlator input.

The selection of sample times depends on the rate of decay of the correlation function. For a single exponential (monodisperse, rigid spheres), using a correlator capable of providing an experimental baseline, the measured correlation function should drop to about $\exp(-4)$ for self-beating experiments (20). The condition becomes $\Delta t = 2/M\Gamma$. Table 2

---

* The cost of current electronics limits n to 1 pulse/100 nsec. Full 4-bit multiplication occurs for sample times above 1.5 nsec. This is not a serious limitation for most particle sizing applications. Faster, single-clipping correlators have been produced, but at much higher cost.

Table 2. Sample Time (μsec) for a 64-Channel Correlator as a Function of Diameter and Scattering Angle[a]

d	θ		
(μm)	30°	90°	150°
2	1560	210	110
0.2	156	21	11
0.02	15.6	2.1	1.1
0.002	1.56	0.21	0.11

[a] Assume that λ_0 = 632.8 nm, 20°C, and H_2O as the suspending medium.

shows the sample time needed for various particle diameters according to the above condition, assuming that a 64-channel correlator is used. To cover the size range, the sample times should be selectable from a few tenths of a microsecond to a few milliseconds.

The measured correlation function approaches the true function as $N \rightarrow \infty$. Also, the total experiment duration ($N\Delta t$) depends on N. If N is too small, the statistical error is large; if N is too large, experiments are inconveniently long. The relative error in the correlation function is a maximum at long times and it may be estimated as $\pm (M'/N)^{1/2}$, where M' is the number of points at which the correlation function is measured and N is the number of sampling intervals.

For a 64-channel correlator, Table 3 shows the number of samples necessary to achieve a predetermined level of error.

With the values in Table 2 and assuming that a relative error in the correlation function of 0.5% is acceptable, the total experiment duration ($N\Delta t$) is shown in Table 4 for various diameters and angles. Except for low angles and large particles, the total time is short.

The sample concentration for optimum results is difficult to calculate *a priori*. It depends on too many variables. The sample should appear clear to slightly turbid. The laser beam in the sample should look like a crisp line, not fuzzy or glowing. For a latex particle of 0.1-μm diameter, c = 20 μg/cm^3 (0.002% wt/vol) or less is sufficient. Since scattering is proportional to NM^2 (here N is the number of particles) or cM, and M is proportional to d^3, the concentration for larger particles is much less. However, for particles larger than the wavelength, intraparticle interference may require larger concentrations at angles of 90° or greater. When concentration is too high and significant interparticle interactions

Table 3. Number of Sampling Intervals N Necessary to Achieve the Given Relative Error, Assuming a 64-Channel Correlator

N	Relative Error (%)
2.6×10^4	5
1.6×10^5	2
6.4×10^5	1
2.6×10^6	0.5
1.6×10^7	0.2
6.4×10^7	0.1

exist, D_T and the calculated size are a function of concentration. In this case, experiments as a function of concentration should be run with extrapolation to zero concentration.

Only coherently detected light contributes to the exponential decay. Detection is optimum when approximately one coherence area is detected (20). The coherence area is defined as the area over which the phase of the scattered light is correlated. For the two-pinhole arrangement shown in Figure 4, the criterion becomes

$$\frac{d_{p1}d_{p2}}{\lambda_0 L} = 1 \tag{11}$$

where d_{p1} and d_{p2} are the diameters of pinholes P1 and P2, respectively,

Table 4. Total Experiment Duration (sec) for Various Diameters and Scattering Angles at the 0.5% Relative Error Level in a 64-Channel Autocorrelation Function

d (μm)	θ 30°	90°	150°
2	4056	546	286
.2	406	55	29
.02	41	6	3
.002	4	.6	.3

and L is the distance between them; d_{p2} is calculated from Equation 11. With high intensities, it may be necessary to select a smaller diameter; for weak scattering situations, it may be necessary to select a larger diameter to obtain results in a reasonable time.

5. DATA ANALYSIS AND INTERPRETATION

The simplest case involves single exponential decay for a monodisperse sample with no interference from dust or flare light. The monitor channels (total pulses and total samples) are used to calculate the infinite-time baseline for normalization.

After normalizing the autocorrelation function with the baseline and subtracting 1, the remaining part, $b \exp(-2\Gamma m\Delta t)$, is fit using a least-squares technique. The value of b typically varies from 0.2 to 0.8, depending upon the optical configuration; b is a maximum when the coherence criterion of Equation 11 is met. The diffusion coefficient D_T is calculated from Γ, and the particle diameter is calculated from D_T. Typical results of experiments on latex spheres carried out more than 10 years ago show that the technique is as reliable as an electron microscope on monodisperse samples of latex spheres (21).

The size calculated from Equation 9 is a hydrodynamic diameter. It may be larger than the diameter of the dry particle because of a layer of solvent molecules, associated ions for charged particles, or surfactant molecules that may have been attached to the particle surface to prevent aggregation. In most cases, these layers add a negligible amount to the diameter, except for the very smallest sizes measurable.

The results for monodisperse samples can be compared directly with other techniques, since there is only one "average" diameter. For nonspherical particles, there are two choices: either the results can be interpreted in terms of an equivalent sphere diameter equal to the Stokes–Einstein hydrodynamic diameter or, for simple, nonspherical shapes, the diffusion coefficient can be interpreted in terms of ellipsoids of revolution (22). This requires either independent knowledge of one dimension or ratio of dimensions or angular measurements from which, in favorable cases, both dimensions can be calculated (23).

For polydisperse samples, the interpretation of data is considerably more difficult. Since the technique does not involve the counting of single particles, size distribution information must be obtained from the deconvolution of the sum over all the single exponentials contributing to the measured autocorrelation function. The deconvolution of summed exponentials is difficult. The problem may be summarized by the following equation

$$|g^{(1)}(\tau)| = \int_0^\infty G(\Gamma)\exp(-\Gamma\tau)\,d\Gamma \tag{12}$$

where $|g^{(1)}(\tau)|$, the first-order autocorrelation function, is the experimentally determined quantity (after baseline normalization, subtraction of unity, and taking the square root of the remaining self-beat function) and $G(\Gamma)$ represents the distribution of linewidths owing to the distribution of sizes (24).

A numerical technique has been applied to the solution of Equation 12 by Chu et al. (25). The result is a histogram-like distribution of $G(\Gamma)$, the first step necessary for particle size distribution information. However, the technique, in its present form, does not always lead to converging results. Following the work of McWhirter and Pike (26) and McWhirter (27) on the Laplace transform of Equation 12, Ostrowsky et al. (28) demonstrated that $G(\Gamma)$ can be calculated at exponentially spaced sample times. Provencher et al. (29) also developed an extremely effective inversion scheme for use with the more usual, equally-spaced sample times.

The approaches described above show that, from typical experimental measurements of the correlation function, only a few parameters of the distribution can be calculated. The present work on these advanced techniques is focusing on the separation of only two peaks in a size distribution, the ratio of which is at least 2 or 3 to 1 in the size. Thus, for the present, PCS measurements must be considered as relatively insensitive to the particular size distribution; except for monodisperse samples, for which the technique works extremely well, only a few moments of the distribution can be obtained.

The most widely used and simplest data analysis technique to apply is the method of cumulants (24, 30). In practice, two moments of the diffusion coefficient distribution are obtained. The method proceeds by expanding Equation 12 about an average linewidth designated $\bar{\Gamma}$. The result is

$$\ln[b^{1/2}|g^{(1)}(\tau)|] = \ln(b^{1/2}) - \bar{\Gamma}\tau + \frac{\mu_2\tau^2}{2} - \frac{\mu_3\tau^3}{6} + \cdots \tag{13}$$

where Γ, μ_2, and so on are moments of the linewidth distribution.* Thus,

* In a cumulant expansion, the moments and cumulants are identical up to the third cumulant.

the normalized correlation function is fit to a polynomial in τ. In practice, only the first two moments are obtained with certainty, and care must be taken to limit the range of τ such that higher-order terms are negligible. This can be difficult for broad distributions.

This technique has the advantage that no assumption about the form of the distribution is necessary and, under the correct experimental conditions described below, the moments are well-defined and useful parameters.

Brown et al. (24) showed that

$$\bar{\Gamma} = \bar{D}\, q^2 \tag{14}$$

where $\bar{D}$ is given by

$$\bar{D} = \frac{\sum N_i M_i^2 P_i(\theta) D_i}{\sum N_i M_i^2 P_i(\theta)} \tag{15}$$

where NM^2P is the time-averaged intensity weighting factor from Equation 1. The sum is over all the particles contributing to the scattering, and it is assumed that the measurements have been made or will be extrapolated into the low concentration regime. If not, the interparticle interference $B(c)$ must be included in Equation 15. It is also assumed that the optical constant K in Equation 1 is the same for all particles, independent of size, and K includes the refractive index increment (change of solution refractive index with particle concentration). Hence, the assumption of constant K means assuming a sample of homogeneous composition independent of size.

For Rayleigh scatterers ($d \ll \lambda$) and for all particles where measurements have been extrapolated to zero-angle, $P(\theta) \rightarrow 1$ and

$$\bar{D} = \bar{D}_z = \frac{\sum N_i M_i^2 D_i}{\sum N_i M_i^2} \tag{16}$$

where $\bar{D}_z$ is the well-defined, z-average diffusion coefficient.

Thus, for spheres, where $M^2 \alpha d^6$, the particle average obtained is

$$\left(\frac{\bar{1}}{d}\right)_z = \frac{\sum N_i d_i^5}{\sum N_i d_i^6} \tag{17}$$

the inverse, z-average diameter.

The second moment in the cumulant analysis, μ_2, yields (after extrapolation to zero concentration and zero angle)

$$\mu_2 = \overline{D_z^2} - \overline{D}_z^2 \tag{18}$$

which is the variance of the z-average diffusion coefficient distribution. The first term can be related to the inverse square, z-average diameter, $(1/d^2)_z$, for spherical particles.

A convenient polydispersity index is defined as

$$Q = \frac{\mu_2}{\overline{\Gamma}^2} = \frac{\overline{D_z^2} - \overline{D}_z^2}{\overline{D}_z^2} \tag{19}$$

In contrast to the ensemble averaging technique represented by PCS measurements, the data obtained from a single particle counter technique can be used to calculate the number-average diameter, d_n,

$$\overline{d}_n = \frac{\sum N_i d_i}{\sum N_i} \tag{20}$$

and, in principle, all other moments. In practice, the higher moments contain progressively more error. Some single particle techniques are sensitive to the weight of particles in a given size class. The data from these measurements can be used to calculate the weight-average diameter given by

$$\overline{d}_w = \frac{\sum w_i d_i}{\sum w_i} = \frac{\sum N_i M_i d_i}{\sum N_i M_i} = \frac{\sum N_i d_i^4}{\sum N_i d_i^3} \tag{21}$$

and, in principle, all other moments.

Thus, the two size parameters from a PCS measurement using cumulant analysis are not directly comparable with the moments obtained from other size distribution measurements. This is the result of two effects: the inverse relationship between diffusion and size and the z-average weighting from the intensity. Furthermore, unless measurements have been extrapolated to zero angle and zero concentration, the apparent size parameters obtained are angle and concentration dependent.

For quality control measurements, the exact interpretation of data is not usually of primary importance. In this case, measurement at one angle, usually 90°, is sufficient to establish an apparent average size and apparent measure of the width of the size distribution. Monitoring changes in these two parameters may be all that is required. Although it can be misleading, it is sometimes assumed at this point in the analysis that a two-parameter, unimodal function describes the particle size distribution. The two, model-dependent parameters can be calculated from the first two cumulants. From this information, any moment of the size distribution can be calculated, as well as tabular values and graphical displays of the histogram and cumulative size distributions. Such an assumption is only warranted under restricted conditions, when transformation of the basic data is useful for visual presentation of the results or when a narrow, unimodal distribution is known to be present. The results should be appended with appropriate precautions. Detailed deductions based on the shape of the size distribution calculated under these assumptions can be very misleading, especially for broad distributions. The breadth of the distributions is reflected in the value of the second cumulant, and the uncertainty in μ_2 is particularly sensitive to dust and flare.

As mentioned in Section 3, flare light and dust are problems. They can seriously complicate data interpretation (31). The intensity from flare is not time-dependent; however, it can mix with the scattered light signal (local oscillator effect), giving rise to an additional term in Equation 7 of the form $\exp(-\Gamma\tau)$. The effect depends on the ratio of the local oscillator and the scattered signal strengths. Upward curvature of the correlation function is obtained with an apparent increase in average particle size and a large increase in the polydispersity, when the measured function is force-fit to a self-beating model. Sometimes, a strong local oscillator is introduced intentionally. When done properly, the self-beat term is negligible and data analysis proceeds from assuming that only the local oscillator term is important. However, it is difficult to control the relative strengths of the local oscillator and scattered light signal for a distribution of sizes, especially for angular measurements. For most measurements, a carefully constructed experimental configuration that excludes flare is recommended.

Once excluded by design, flare is a nonrecurrent problem. Dust, however, may be introduced every time a new sample is measured. Dust is defined as any very large contaminant that is not part of the intended sample. Being larger, its contribution to the measured function increases with decreasing angle. If the dust size is very much greater than the particle size, then the self-beat term from the dust is flat over the time range characterizing the sample of interest. In this case, a baseline calculated from several channels, which have been delayed by at least $12/\overline{\Gamma}$, is a better choice than the infinite-time baseline. It is then assumed that no large scatterers of interest are contributing significant information to the delay channels. Dust can also act as a local oscillator, causing an apparent increase in average size and polydispersity. This effect is in addition to the actual increase in these parameters when the dust is considered as part of the sample.

Since the light scattered from dust is angular dependent, higher angles are preferred. However, it was shown previously that measurements and extrapolation to the low angle regime are often necessary for the correct data interpretation. Filtering all liquids, especially water, and thoroughly cleaning all sample cells, pipettes, syringes, and so on are often necessary. A comparison between baselines, using delay and monitor channels, can be used to accept or reject data on the basis of dust contamination. Sometimes, a series of short measurements, with dust rejection, are combined.

Time-averaged intensity measurements are often plagued by dust. Its contribution is a systematic increase in intensity. There is no way to remove it by subtraction. In PCS, dust is also a problem. However, in some cases, its effects can be removed by subtraction of the delayed baseline. In this sense, PCS data are less sensitive to dust than pure intensity measurements.

6. SUMMARY

In Section 1, it was mentioned that no single technique can perform all particle sizing requirements; PCS is no exception. It has several *disadvantages*:

1. It does not, in general, produce a high resolution histogram of the size distribution.
2. Like other nonimaging techniques, an equivalent sphere diameter is usually, although not always, assumed. That is, shape information is not easily obtained.
3. When proper measurements are made, the parameters of the size distribution most often obtained—the inverse z-average moments—are not the usually reported parameters of a size distribution.
4. Although less sensitive to "dust" than intensity measurements, it can still make measurement and interpretation difficult.

The PCS method also has many *advantages* to consider:

1. Measurements are made in seconds to minutes.

2. The technique is absolute. Calibration with a known size distribution is not necessary.

3. Very small quantities of sample can be used.

4. Any suitable suspending liquid can be used, provided that it is nonabsorbing, relatively clear, and not too viscous.

5. The technique is applicable from about 0.002 to several microns.

6. Instrumentation is commercially available for both research and quality control measurements, with automation including data analysis.

7. Although the interpretation of particle size is least ambiguous with a narrow distribution, an effective diameter and polydispersity index are measureable, even with broad distributions.

New data analysis techniques, which are currently under investigation (25–29), promise to improve the amount of particle sizing information available on a routine basis from PCS measurements.

NOTATION

b	Experimental constant.
B	Concentration factor.
c	Particle concentration.
$C(\tau)$	Photon correlation function.
d	Particle diameter.
$\overline{d}_n$	Number average particle diameter.
dp_1 and dp_2	Diameter of pinholes.
$\overline{d}_w$	Weight average particle diameter.
D_T	Translational diffusion coefficient.
$\overline{D}_z$	Z-average diffusion coefficient.
$\mathbf{E}_I$	Electric field vector.
$\mid g^{(1)}(\tau) \mid$	First-order correlation function.
$G(\Gamma)$	Linewidth distribution.
$\langle I \rangle$	Scattered light intensity.
$\langle I(0)I(\tau) \rangle$	Autocorrelation function of the scattered light intensity.
$\langle I_s(q) \rangle$	Time-averaged scattered intensity from suspended particles.
k	Boltzmann constant.

$\mathbf{k}_I$	Incident wave vector.
$\mathbf{k}_s$	Scattering wave vector.
K	Optical constant.
L	Distance between pinholes.
M	Particle mass.
M'	Number of points at which the correlation function is measured.
n	Index of refraction of suspending media or number of pulses or photons.
$n(t)$	Number of pulses registered between time $t - \tau$ and t.
N	Number of particles contributing to the scattering, number of sampling intervals, or number of sample time intervals of duration Δt.
P	Particle form factor.
q	Wave vector amplitude of the scattering fluctuation.
Q	Polydispersity index.
$\mathbf{r}$	Vector describing spatial position.
t	Initial time for measuring light intensity fluctuations.
Δt	Sample or delay time.
T	Total averaging time or absolute temperature.
Γ	Linewidth of scattered light.
$\bar{\Gamma}$	Average linewidth.
η	Viscosity of the medium in which particles are suspended.
θ	Scattered angle from incident beam.
λ_0	Wavelength of light in vacuum.
λ_q	Wavelength of fluctuation causing scattering.
μ_n	Moments of linewidth distribution.
τ	Delayed time for measuring light intensity fluctuations.

REFERENCES

1. T. Allen, *Particle Size Measurement*, 3rd ed., Chapman & Hall, New York, 1981.
2. B. Chu, *Laser Light Scattering*, Academic Press, New York, 1974.
3. B. J. Berne and R. Pecora, *Dynamic Light Scattering with Applications to Chemistry, Biology and Physics*, Wiley-Interscience, New York, 1976.
4. H. Z. Cummins and E. R. Pike (Eds.), *Photon Correlation and Light Beating Spectroscopy*, Plenum Press, New York, 1974.

5. H. Z. Cummins and E. R. Pike (Eds.), *Photon Correlation Spectroscopy and Velocimetry*, Plenum Press, New York, 1977.

6. V. Degiorgio, M. Corti, and M. Giglio (Eds.), *Light Scattering in Liquids and Macromolecular Solutions*, Plenum Press, New York, 1980.

7. S. H. Chen, B. Chu, and R. Nossal (Eds.), *Scattering Techniques Applied to Supramolecular and Nonequilibrium Systems*, Plenum Press, New York, 1981.

8. R. Foord, E. Jakeman, C. J. Oliver, E. R. Pike, R. J. Blagrove, E. Wood, and A. R. Peacocke, *Nature,* **227,** 242 (1970).

9. H. Z. Cummins, "Applications of Light Beating Spectroscopy to Biology," in ref. 4, pp. 312–313.

10. J. Tyndall, *Phil. Mag.,* **37,** 384 (1869).

11. J. W. Strutt, *Phil. Mag.,* **41,** 447 (1871).

12. Lord Rayleigh, *Phil. Mag.,* **12,** 81 (1881).

13. M. Smoluchowski, *Ann. Phys.,* **25,** 205 (1908).

14. A. Einstein, *Ann. Phys.,* **33,** 1275 (1910).

15. H. C. Van de Hulst, *Light Scattering by Small Particles*, John Wiley and Sons, New York, 1957.

16. C. Tanford, *Physical Chemistry of Macromolecules*, John Wiley and Sons, New York, 1961, Chapter 5.

17. R. Pecora, *J. Chem. Phys.,* **40,** 1604 (1964).

18. H. Z. Cummins, N. Knable, and Y. Yeh, *Phys. Rev. Lett.,* **12,** 150 (1964).

19. E. Jakeman and E. R. Pike, *J. Phys* A1, **2,** 411 (1969).

20. C. J. Oliver, "Correlation Techniques," in ref. 4, pp. 151–223.

21. S. P. Lee, W. Tscharnuter, and B. Chu, *J. Polym. Sci.,* **10,** 2453 (1972).

22. C. Tanford, ref. 16, p. 327.

23. H. Z. Cummins, F. D. Carlson, T. J. Herbert, and G. Woods, *Biophys. J.,* **9,** 518 (1969).

24. J. C. Brown, P. N. Pusey, and R. Dietz, *J. Chem. Phys.,* **62,** 1136 (1975).

25. B. Chu, Es. Gulari, and Er. Gulari, *Phys. Scr.,* **19,** 476 (1979).

26. J. G. McWhirter, E. R. Pike, *J. Phys. A: Math. Gen.,* **11,** 1729 (1978).

27. J. G. McWhirter, *Opt. Acta,* **27,** 83 (1980).

28. N. Ostrowsky, D. Sornett, P. Parker, and E. R. Pike, *Opt. Acta,* **28,** 1059 (1981).

29. S. W. Provencher, J. Hendrix, L. De Maeyer, and N. Paulussen, *J. Chem. Phys.,* **69,** 4273 (1978); S. W. Provencher, *Macromol. Chem.,* **180,** 201 (1979).

30. D. E. Koppel, *J. Chem. Phys.,* **57,** 4814 (1972).

31. C. J. Oliver, in ref. 7, p. 137.

APPLICATION OF PHOTON CORRELATION FUNCTION PROFILE ANALYSIS TO MOLECULAR WEIGHT DISTRIBUTIONS OF POLYMERS IN SOLUTION

BEN CHU and ART DiNAPOLI

Chemistry Department
State University of New York
Stony Brook, New York

1. INTRODUCTION

Several methods of data analysis for extracting information on the distribution of translational diffusion coefficients of a solution of polydisperse macromolecules from the intensity autocorrelation function of the scattered light have been reported (1–5). In particular, the cumulants expansion technique (1) can be used to estimate the first and second moments of the linewidth distribution, $G(\Gamma)$, which is directly related to the distribution of diffusion coefficients, $G(D)$. Although the first and second moments of $G(\Gamma)$ often provide sufficient information on the average size and polydispersity index (or variance) of the macromolecules in solution, at times we wish to determine an approximate $G(D)$ to obtain an estimate of the molecular weight distribution, $f(M)$. To this end, the histogram approximation (2, 3) and the Pearsons approach are feasible methods that permit us to approximate $G(\Gamma)$. The histogram technique approximates $G(\Gamma)$ as a series of descrete segments in Γ-space; the Pearsons approach assumes that $G(\Gamma)$ can be approximated by a distribution function written as a Pearsons Type I, five-parameter equation. In this chapter, the transformation procedure presented previously (4, 5) is extended to a synthetic polymer, polyacrylamide (PAM), that has a moderately broad molecular weight distribution with $M_w/M_n \simeq 2$. PAM was studied recently in some detail (6, 7), allowing the transformation to be performed quantitatively. Finally, we provide a brief summary of other inversion techniques for the time correlation function (or the power spectrum). Some algorithms converge faster than others, but all the techniques provide a good approximation for $G(\Gamma)$ without an *a priori* assumption on the form of $G(\Gamma)$. On the other hand, the superposition of exponentials, the finite time range of the correlation function (or the power spectrum), and the experimental signal-to-noise ratio usually limit the resolution of $G(\Gamma)$ to unimodal and bimodal distributions. Fortunately, polymer molecular weight distributions are mostly unimodal in form, are sometimes bimodal, but are rarely beyond trimodal. Therefore, we can use correlation function profile analysis to approximate $G(\Gamma)$, which is related to $f(M)$.

2. TRANSFORMATION METHOD

The first-order electric field correlation function for a polydisperse system can be written as

$$g^{(1)}(\tau) = \int_0^\infty G'(\Gamma) \exp(-\Gamma\tau)d\Gamma \tag{1}$$

where τ is the delay time (on the order of microseconds) and $G'(\Gamma)$ denotes a normalized linewidth distribution. We can also write

$$g^{(1)}(\tau) = \frac{\langle E^*(0)E(\tau)\rangle}{\langle |E|^2\rangle} \tag{2}$$

where $E(\tau)$ is the magnitude of the electric field vector and the bracket indicates a statistical average. In the limit of $\tau \to 0$, Equations 1 and 2 lead to the result

$$\langle I(\theta)\rangle \propto \int_0^\infty G(\Gamma)d\Gamma \tag{3}$$

which shows that the area under the $G(\Gamma)$ curve is proportional to $\langle I(\theta)\rangle$, the time averaged, total intensity at the scattering angle, θ.

We can further show (8) that the intensity can be written as

$$\langle I(\theta) \rangle = K' \sum_i M_i C_i P(\theta, M_i) \tag{4}$$

where K' represents a collection of constants depending on the experimental optics, the incident intensity, wavelength, and polarization, and the polarizability per unit mass of the polymer; C_i is the concentration (mass per volume) of species with a molecular weight M_i; and $P(\theta, M_i)$ is the particle scattering factor, representing the spatial configuration of the polymer with M_i. Equation 4 is valid only when $A_2 C \ll 1/M_w$, where A_2 is the second virial coefficient, M_w is the weight average molecular weight, and $C(\equiv \sum_i C_i)$ is the total concentration. By writing $C_i = f(M_i)\Delta M_i/V$, where $f(M_i)$ is the weighted molecular weight distribution, ΔM_i is an increment along the M-axis such that $f(M_i)\Delta M_i$ represents a small area under the molecular weight distribution curve and V is a unit volume, we can relate $f(M)$ and $G(\Gamma)$ as

$$\int_0^\infty G(\Gamma)d\Gamma = K'' \int_0^\infty f(M)MP(\theta, M)\, dM \tag{5}$$

with the proportionality constant of Equation 3 being absorbed into K' and rewritten as K''. The transformation process is performed in a piecewise manner, equating small areas under the $G(\Gamma)$ curve with areas under the $f(M)$ curve, using Equation 6 as a basis for the process:

$$G(\overline{\Gamma}_i)\Delta\Gamma_i = f(\overline{M}_i)\overline{M}_i P(\theta, \overline{M}_i)\Delta M_i \tag{6}$$

In using the linewidth distribution deduced from the histogram analysis, the step width in Γ-space, $\Delta\Gamma_i$, is equal to the width of the histogram, and $\overline{\Gamma}_i$ and $\overline{M}_i$ are the center of the step in Γ-space and in molecular weight space, respectively. In this case, $\Delta\Gamma_i$ is equal for each step, although in principle logarithmically spaced end points should give better results (9). The Pearsons approach yields a continuous distribution curve and, therefore, the step width can be selected arbitrarily. We normally use a narrower step size at the low-frequency end of the Γ-distribution, which is more susceptible to distortion in the transformation process.

To evaluate the parameters on the right side of Equation 6, and ultimately $f(\overline{M}_i)$, we must relate the linewidth, Γ, to the molecular weight, M. At the outset, the assumption is made that the line broadening is solely due to the translational motion of the polymer. This is by no means a trivial or widely applicable assumption, as we discuss further, but having made it, we may relate the linewidth to the diffusion coefficient as

$$\Gamma(\theta, C) = D(\theta, C)K^2 \tag{7}$$

with $K = 4\pi n_0 \sin(\theta/2)/\lambda_0$, where n_0 and λ_0 are, respectively, the refractive index of the solvent and the wavelength *in vacuo* of the incident light. The concentration dependence is traditionally (10) taken into account as

$$D(\theta, C) \cong D_0(\theta)(1 + k_D C) \tag{8}$$

where D_0 is the translational diffusion coefficient at infinite dilution and k_D is the diffusion second virial coefficient. Higher-order terms in C have been neglected. We rely on a simple plot of $D(\theta, C)$ vs. C to obtain k_D in the linear region, rather than on theoretical expressions (11). For a monodisperse solution, the angular dependence of the diffusion coefficient is strictly corrected for by K^2, but for even moderately broad distributions, we see deviations from the K^2-scaling, since the particle scattering factor is not the same for each molecular-weight component. This problem, pointed out by Caroline (12), has been circumvented in the approach of Raczek and Meyerhoff (13) and previously considered by Burchard (14). Since we demand low concentrations throughout the transformation process, the term $k_D C$ is quite small and little error is introduced by using the value of k_D for PAM in water obtained in the limit of $\theta \rightarrow 0$, as shown in Figure 1.

For the remainder of this chapter, we drop the θ argument from D_0 and write the relationship between D_0 and M as

$$D_0 = k_T M^{-b} \tag{9}$$

The value of k_T depends on the details of the polymer–solvent system and cannot be predicted. The exponent, b, is known to be 1/2 for random coils in a theta solution at the theta temperature, both theoretically and experimentally. The values of b for random coils in good solvents, rods, and hard spheres are known theoretically, but experimental verification is lacking. The approach described here will determine both k_T and b experimentally.

By combining Equations 7, 8, and 9, we arrive at the desired result:

$$M = \left[\frac{\Gamma}{k_T K^2(1 + k_D C)} \right]^{-1/b} \tag{10}$$

Equation 10 is used to compute M in Equation 6 at the endpoints and center of each segment in Γ-space. The absolute value of the difference

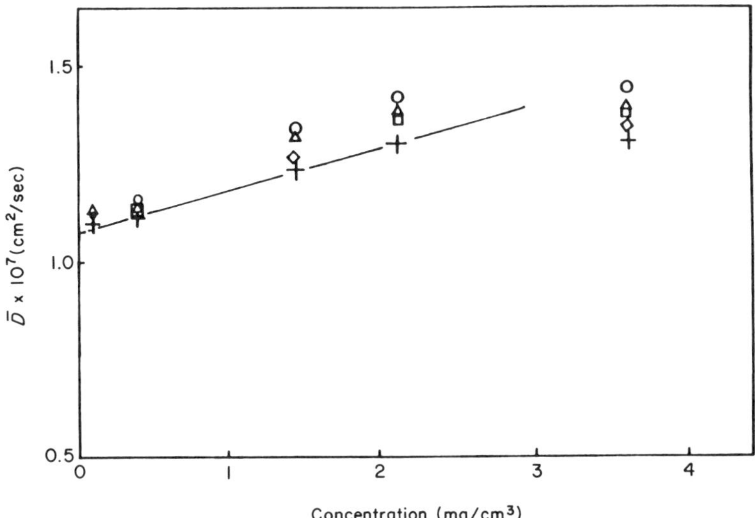

Figure 1.   The diffusion second virial coefficient, k_D (equation 8), is calculated graphically for PAM-B in water. At different angles [open circles (○) are $\theta = 90°$; triangles (△) are $\theta = 75°$; squares (□) are $\theta = 60°$; diamonds (◇) are $\theta = 45°$; and crosses (†) indicate the data extrapolated to $\theta = 0°$] different slopes relating to the molecular weight dependence of k_D are measured. Results from the zero scattering angle line are $D_0 = 1.08 \times 10^{-7}$ cm^2/sec, $k_D(\theta \to 0) = 100$ cm^3/g.

in the computed molecular weights at the endpoints is ΔM_i; the value of $\overline{M}_i$ is computed directly from $\overline{\Gamma}_i$. The nonlinear nature of the transformation produces segments in M-space that become quite broad in the high molecular weight regime (i.e., the low-frequency end of Γ-space). For this reason, smaller spacings at the low frequencies in Γ-space for the Pearsons approach are used.

In the transformation, we need to evaluate the particle scattering factor, $P(\theta, M)$, as a function of molecular weight. For random coil molecules, $P(X, M)$ can be expressed as

$$P(X, M) = \frac{2}{X^2} [\exp(-X) - 1 + X] \tag{11}$$

where $X = K^2 r_g^2$ and r_g is the radius of gyration. Thus, by relating the radius of gyration and molecular weight in an expression analogous to Equation 9,

$$r_g = k'_T M^{b'} \tag{12}$$

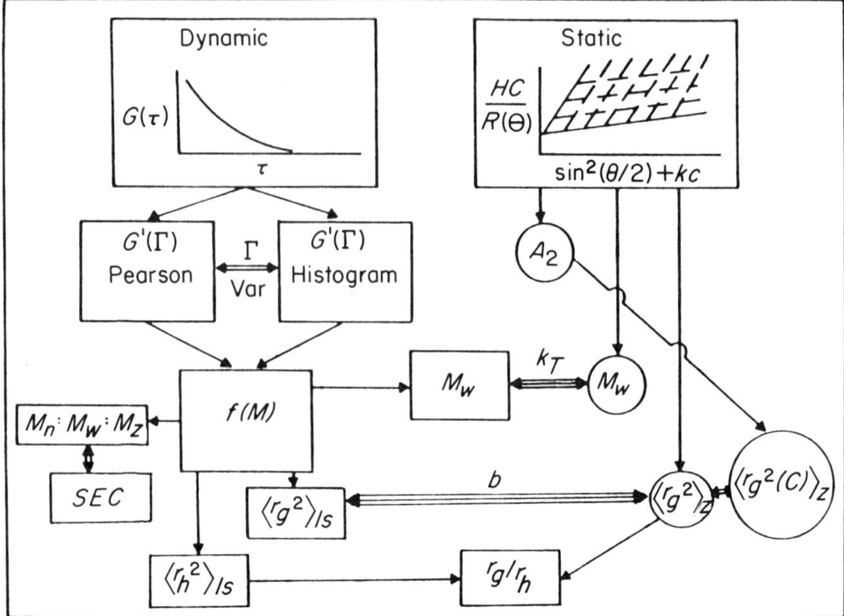

Figure 2. The transformation from a linewidth distribution, $G(\Gamma)$, to a molecular weight distribution, $f(M)$, requires both dynamic and static light scattering data. Single lined arrows indicate the progression from one quantity to another. Double arrows indicate simple comparisons for internal consistency and the two triple arrows indicate that we force the measured and deduced M_w's to be equal through k_T and $\langle r_g^2 \rangle$ to be equal through b.

Equations 11 and 12 are also needed in order to perform the transformation. This is only possible if the polymer configuration is represented by Equation 11 and the parameters k_T' and b' are known. Both conditions are satisfied for polystyrene (4, 15) and PAM (6).

The entire transformation process is outlined in flow chart form, as shown in Figure 2. Both the dynamic and static light scattering data are required. The classical technique for determining the second virial coefficient, A_2, the weight-average molecular weight, M_w, and z-average, concentration independent radius of gyration, $\langle r_g^2 \rangle_z^{1/2}$, is well-known (18) and is not discussed further. The concentration-dependent radius of gyration $\langle r_g^2(C) \rangle_z^{1/2}$ is also determined by straightforward means as a check of the value $\langle r_g^2 \rangle_z^{1/2}$.

The correlation function can be analyzed using moments analysis (1) as well as the histogram (2–4) and Pearsons (2, 4) techniques. Equation 6 is then used with Equation 10 and, in the case of random coils, Equations

11 and 12 to determine $f(M)$. Values of k_T and b are estimated initially. From the deduced $f(M)$, M_w can be computed and compared with the experimental results from a Zimm plot, using integrated scattered intensity measurements. The value of k_T is adjusted until the two M_w values agree. When k_T' and b' are known, the z-average radius of gyration from the deduced $f(M)$ can be evaluated and compared to the measured value. For a meaningful comparison, we must compute the radius of gyration from $f(M)$ as,

$$\langle r_g^2 \rangle_{ls} = \frac{\sum f(M_i) r_{g,i}^2 \, P(\theta, M_i) M_i \, \Delta M_i}{\sum f(M_i) P(\theta, M_i) M_i \, \Delta M_i} \tag{13}$$

which is the z-average, $P(\theta, M)$-weighted radius of gyration, rather than the simple z-average value. Agreement between $\langle r_g^2 \rangle_z^{1/2}$ from the Zimm plot and $\langle r_g^2 \rangle_{ls}^{1/2}$ confirms whether we have used a correct value for b.

3. EXPERIMENTAL TECHNIQUES

Details of the instrumentation and sample preparation have been described elsewhere (17).

An argon–ion laser operating at $\lambda_0 = 488.0$ nm with typical beam output of less than 150 mW was used. Dynamic measurements were performed with a Malvern K7023 96-channel single-clipped correlator that was modified to allow direct measurements of the baseline.

Polyacrylamide (PAM-A) was purchased from Polyscience (Cat. #8248, Lot 93-3, Polyscience Inc., Warrington, PA.). In addition, a second sample (PAM-B) was prepared specifically for this study by C. Y. Cha (Calgon Corporation, Calgon Center, Box 1346, Pittsburgh, PA. 15230). For PAM-B, the polymerization was stopped after 10% yield to obtain an intermediate molecular weight distribution of the polymer.

A 1 wt% stock solution was prepared using solvent that had been filtered through a 0.05-μm nominal pore diameter membrane. The light scattering cells were rinsed in an acetone still prior to use to remove dust.

4. RESULTS

Measured static properties of the polymers are listed in Table 1. We used simulated data to check the internal consistence and logic of the computer programs. A series of molecular weight distributions as generated using

Table 1.   Static Parameters of Polymers Used in This Study

Sample	Solvent	M_w (g/mole) $\times 10^{-5}$	$\langle r_g^2 \rangle_z^{1/2}$ (Å)	A_2 (mole cm^3/g^2)
PAM-A	0.5 N NaNO$_3$	2.23 $\pm$ 0.09	250 $\pm$ 25	4.3 $\pm$ 0.2
PAM-A	H$_2$O	2.6 $\pm$ 0.1	320 $\pm$ 20	3.9 $\pm$ 0.2
PAM-B	H$_2$O	4.6 $\pm$ 0.1	415 $\pm$ 20	3.9 $\pm$ 0.2

a Schultz distribution, from which M_n, M_w, and M_z were computed. The measured parameters for polystyrene at the theta condition (15) were used to determine the z-average, $P(\theta, M)$-weighted radius of gyration and the diffusion coefficient. From $f(M)$, a correlation function with added error was computed and analyzed to deduce $G(\Gamma)$. We then determined the molecular weight distribution $f(M)$ from $G(\Gamma)$.

In Table 2, parameters computed from the $f(M)$ deduced through the $g^{(1)}(\tau) \rightarrow G(\Gamma) \rightarrow f(M)$ transformation are compared with those computed directly from the input $f(M)$. The simulated data sets are the same as those used in reference (18). We also measured the time correlation function for NBS-705A polystyrene in cyclohexane at 35°C and two scattering angles. The linewidth distribution was determined using the Pearsons approach and was then transformed to the molecular weight distribution with the parameters given in reference (15). These results are tabulated in Table 3.

The transformation of the PAM linewidth distributions is more complicated because k_T and b of Equation 9 are unknown. For a polymer in a good solvent, such as PAM in water, the exponent of the static scaling expression, Equation 12, is expected to be 0.60. Fractionation of a broad PAM distribution and consequent light scattering studies by Francois et al. (6) (Equation 14a) and Klein and Conrad (7) (Equation 14b) show that

$$\langle r_g^2 \rangle_z^{1/2} = 7.49 \times 10^{-2} M_w^{0.64} \quad (\text{Å}) \tag{14a}$$

$$\langle r_g^2 \rangle_z^{1/2} = 1.86 \times 10^{-1} M_w^{0.59} \quad (\text{Å}) \tag{14b}$$

Although Equation 14b agrees more closely to the theoretical value, we have used Equation 14a to compute $\langle r_g^2 \rangle_{ls}^{1/2}$ as well as $(Kr_g)^2$, which is necessary for the computation of $P(\theta, M)$, according to Equation 12. It should be noted that Equation 12 does not specify any averages, whereas both Equations 14a and 14b do. It can be shown easily (17) that polydispersity, even of relatively narrow fractions, will distort the value of b'.

Table 2. Comparison of Parameters Computed Directly from a Generated $f(M)$ and Those Computed from the Deduced $f(M)^a$

Data Set[b]	M_w (g/mole) $\times 10^5$	Directly from $f(M)$ $\langle r_g^2 \rangle_{ls}^{1/2}$ (Å)	M_w/M_n	M_z/M_w	After Transformation (Pearsons, Histogram) $\langle r_g^2 \rangle_{ls}^{1/2}$ (Å)	$M_w M_n$	M_z/M_w	$k_T \times 10^4$ (cm^2 g$^{1/2}$ mole$^{-1/2}$ sec^{-1})
Theta-VIII	2.00	135	1.02	1.02	140, 136	1.07, 1.06	1.09, 1.06	1.39, 1.37
Theta-I	2.00	144	1.20	1.17	144, 144	1.15, 1.15	1.18, 1.16	1.36, 1.36
Theta-III	1.98	169	2.65	1.64	165, 172	1.35, 1.70	1.63, 1.72	1.29, 1.24
Theta-IV	6.88	264	1.19	1.15	264, 262	1.13, 1.14	1.18, 1.13	1.36, 1.43
Theta-VI	6.88	286	2.61	1.51	280, 282	1.53, 1.64	1.50, 1.48	1.33, 1.35
LT-1[d]	2.66	566	2.07	1.62	548, 562	1.69, 1.85	1.95, 1.68	1.31, 1.31

[a] One more significant figures is presented than would be attainable in a real experiment.

[b] Simulated data were based on parameters of polysytrene in cyclohexane at 35°C, as given by Han and McCracken (15); $g^{(1)}(\tau) = [\Sigma f(M_i)M_i P(X, M_i) \exp(-D_{o,i}K^2\tau)]/\Sigma f(M_i)M_i P(X,M_i)$.

[c] Deviations of the value of M_w are reflected in the parameter k_T, which should agree with the value used to generate the data, 1.37×10^{-4} (cm^2 g$^{1/2}$ mole$^{-1/2}$ sec^{-1}).

[d] LT-1 represents a molecular weight distribution function drawn to simulate a high-pressure liquid chromatography curve and then digitized.

Table 3. Results of Transformation of NBS-705A Polystyrene Standard[a]

	Literature Values	Transformation Results[b]			
		$b' = 0.50$		$b' = 0.49$	
		$\theta = 60°$	$\theta = 90°$	$\theta = 60°$	$\theta = 90°$
$k_T \times 10^4$ (cm^2 g$^{1/2}$/sec mole$^{1/2}$)	1.37	1.49	1.48	1.50	1.50
M_w/M_n	1.07	1.13	1.18	1.13	1.18
M_z/M_w	—	1.19	1.27	1.19	1.27
$\langle r_g^2 \rangle_z^{1/2}$ (Å)	(123)[c]	140	140	120	125
$\langle r_h^2 \rangle^{1/2}/\langle r_g^2 \rangle^{1/2}$	(0.665)[d]	0.60	0.58	0.64	0.65

[a] $C = 0.2$ wt %; $T = 35 \pm 0.01°C$; $M_w = 179{,}000$ g mole^{-1}.
[b] See Equation (12), $k_T' = 3 \times 10^{-9}$.
[c] Estimated value, see text.
[d] Expected theoretical value for $\langle r_h^2 \rangle^{1/2}/\langle r_g^2 \rangle^{1/2}$.

Equation 14a was chosen because the width of the molecular weight distribution of each fraction was clearly specified.

It is not clear whether the static scaling exponent, b', should be identical to the dynamic value, b. Therefore, we must determine b by comparing $\langle r_g^2 \rangle_z^{1/2}$ and $\langle r_g^2 \rangle_{ls}^{1/2}$, as outlined in Figure 2. Tables 4–6 present the transformation results of PAM-A in pure water, PAM-A in 0.5 N NaNO$_3$, and PAM-B in pure water, respectively. An unexpected angular dependence for $\langle r_g^2 \rangle_{ls}^{1/2}$ is found when it is computed in all cases, as shown in

Table 4. Transformation of PAM-A in H$_2$O[a]

θ (degree)	Fit	$k_T \times 10^4$ (cm^2/sec) (mole/g)$^{-.64}$	$\langle r_g^2 \rangle_{ls}^{1/2}$ (Å)	M_w/M_n	M_z/M_w
50	Histogram	4.3	390	1.9	2.4
70		4.9	400	1.9	1.7
85		4.1	355	1.6	2.2
105		4.6	350	1.8	1.5
130		4.4	320	1.6	1.6
50	Pearson	4.0	480	2.1	4.3
70		4.0	395	1.8	3.1
85		4.2	350	2.2	2.2
105		4.0	350	2.0	2.5

[a] $M_w = 2.6 \times 10^5$; $\langle r_g^2 \rangle_z^{1/2} = 320$ Å; $C = 0.1$ wt %; $b = 0.64$.

Table 5. Transformation of PAM-A in 0.5 N NaNO$_3$

θ (degree)	G' $(\Gamma)^a$	$k_T \times 10^4$ (cm^2/sec) (mole/g)$^{-b}$[b]	$\langle r_g^2 \rangle_{ls}^{1/2}$ (Å)	M_w/M_n	M_z/M_w
60	LF	3.6	325	3.4	3.0
60	ENTIRE	4.8	360	3.1	2.7
90	LF	3.6	300	1.5	2.3
90	ENTIRE	4.3	340	2.3	3.1
120	LF	3.25	250	1.4	1.5
120	ENTIRE	4.3	300	2.0	2.4

[a] Histogram fits were used for $G(\Gamma)$. LF indicates that only the low-frequency peak of the bimodal distribution was transformed. ENTIRE indicates that the entire histogram was transformed. See ref. 19 for more extensive analysis.
[b] In all cases, the exponent $b = 0.64$.

Figures 3 and 4 for PAM-A in water and PAM-B in water, respectively. It is observed, especially in Figure 5a, that the value of $\langle r_g^2 \rangle_{ls}^{1/2}$ converges to the expected value of $\langle r_g^2 \rangle_z^{1/2}$ (see Table 1) at higher scattering angles. The implications of this unusal effect are discussed in Section 5.

5. CONCLUSIONS

Table 2 demonstrates that the transformation is valid for moderately broad, unimodal distributions for which k_T, b, k_T', b', and the form of $P(X, M)$ are known. By means of simulated data, the Pearsons method and the histogram linewidth analysis show essentially the same $f(M)$, as expected for a unimodal distribution. However, for broader distributions ($M_w/M_n \gtrsim 3$), the ratio M_w/M_n cannot be determined accurately. Because light scattering measurements detect primarily the larger species of a solution of polydisperse molecules, our results indicate that biasing is introduced in the analysis through the w- and z-averaging processes.

The well-characterized NBS 705 polystyrene standard was selected for its narrow distribution. The form of the particle scattering factor for random coils, Equation 11, is exact only at the theta condition, with no excluded volume effects. The z-average radius of gyration was not measured by means of a Zimm plot in this case, but was estimated in two ways. Through Equation 12, with $M_w = 1.79 \times 10^5$ g mole^{-1}, we obtained $\langle r_g^2 \rangle^{1/2} = 127$ Å. The ratio of the hydrodynamic radius to the radius of gyration for a Gaussian coil is expected to be 0.665. From the measured diffusion coefficient, which yields $\langle 1/r_h \rangle_z$, 119 Å for $\langle r_g^2 \rangle_z^{1/2}$ is computed.

Table 6.   Transformation of PAM-B[a]

C (wt %)	θ (degrees)	Fit[b]	b	$k_T \times 10^{4c}$ (cm^2/sec) (mole/g)$^{-b}$	$\langle r_g^2 \rangle_{ls}^{1/2}$ (Å)[c]	M_w/M_n	M_z/M_w
0.04	45			5.29[c]	475	1.5	2.2
	60	H	0.65	5.23	460	1.6	2.1
	90			4.77	460	1.9	3.2
0.14	45	H		5.41	470	1.7	1.9
	60	H		5.43	440	1.6	1.8
	75	H	0.65	5.43	420	1.7	1.9
	90	H		5.38	420	1.8	1.8
	105	H		5.37	410	1.7	2.9
	45	P		5.23	515	1.5	2.1
	60	P		5.26	460	1.5	2.1
	75	P	0.65	5.25	430	1.5	1.9
	90	P		5.21	430	1.6	2.0
	105	P		5.21	420	1.8	1.8
0.21	45	H	0.60	2.64	651	2.0	3.7
	75	H	0.60	2.74	496	2.1	2.5
	90	H	0.60	1.58	431	1.7	2.0
	45	H	0.65	5.07	596	1.8	3.0
	75	H	0.65	5.26	469	1.9	2.2
	90	H	0.65	3.02	413	1.6	1.8
	45	H	0.68	7.48	569	1.7	2.7
	75	H	0.68	7.77	454	1.8	2.1
	90	H	0.68	4.47	405	1.5	1.7
0.36	45	H		0.89	640	1.9	3.8
	60	H		4.41	580	1.9	4.0
	75	H	0.65	4.50	520	2.0	3.1
	90	H		4.40	495	2.2	3.2
	100	H		4.60	460	2.1	2.5

[a] $M_w = 4.6 \times 10^5$; $\langle r_g^2 \rangle_z^{1/2} = 415$ Å.

[b] H—Histogram fit; P—Pearsons fit.

[c] Three significant figures shown for comparison only, not as an indication of accuracy.

From Table 3, we see that if the transformation is made using the values of b, b', k_T, and k_T' in reference (15), we obtain a value of $\langle r_g^2 \rangle_{ls}^{1/2}$, which does not agree with the expected value of about 123 Å. Furthermore, the ratio r_h/r_g, is too small for a random coil. By changing b' from the anticipated value of $\frac{1}{2}$ to 0.49, more reasonable results are obtained. The analysis cannot be taken too seriously, since $\langle r_g^2 \rangle_z^{1/2}$ has not been meas-

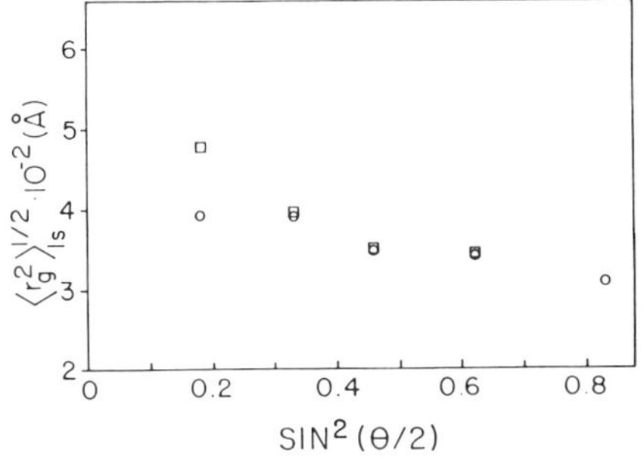

Figure 3. Transformation results of PAM-A in water (0.1 wt %). Pearson fits are represented by (□) and (○) represents histogram fits. In both cases, $b = 0.64$. The expected radius of gyration is 320 Å.

ured, but we wish to point out that the transformation procedure may be used to estimate the exponents of the static and/or dynamic scaling laws.

It is important to note that the transformation of polystyrene at two scattering angles gives the same results. This is in contrast to the results of PAM, as seen in Figures 3 and 4. The deviation between $\langle r_g^2 \rangle_{ls}^{1/2}$ and $\langle r_g^2 \rangle_z^{1/2}$ is particularly troublesome, since it is observed at the lower scattering angles. As θ approaches zero, dynamical contributions to $G(\Gamma)$ from sources other than translational diffusion disappear. For this reason, we would expect any deviations to occur at higher scattering angles. As θ decreases, the scattered intensity from larger molecules increases, when compared to smaller species. Thus, any errors in the process may result in a larger $\langle r_g^2 \rangle_{ls}^{1/2}$, owing to undercompensation of the scattering of large particles and compounded by the z-averaging process. Furthermore, if the $P(X, M)$ weighting of Equation 13 is not included, or if Equation 14b

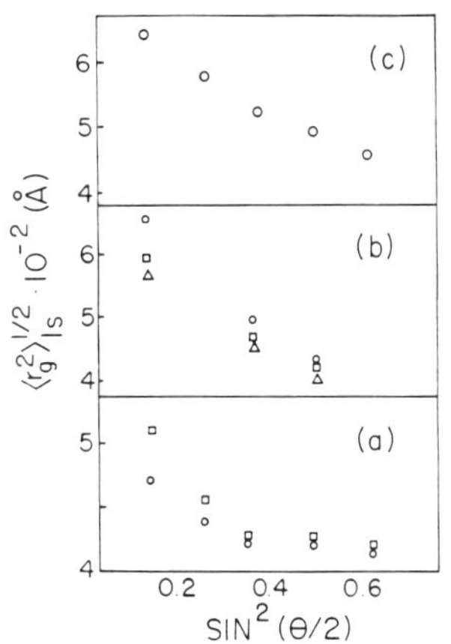

Figure 4. Results from PAM-B in water at $C = 0.14$ wt % (a), $C = 0.21$ wt % (b), and $C = 0.36$ wt % (c). In (a), the histogram (○) and Pearson's approach (□) are compared; (b) shows the effect of varying the value of the exponent b to 0.60 (○), 0.65 (□), and 0.68 (△). The correct value seems to be closer to 0.65. All fits of (b) and (c) are from the histogram approach; (c) shows the transformation results at $C = 0.36$ wt %, $b = 0.65$. The results of these samples and $C = 0.04$ wt % are summarized in Table 6. The expected value, $\langle r_g^2 \rangle_z^{1/2}$, is measured as 415 Å.

rather than 14a is used, we obtain a much higher value of b. In both cases, the angular dependence of $\langle r_g^2 \rangle_{ls}^{1/2}$ is still present.

The histogram of PAM-A in salt solution shows a small, high-frequency peak. The transformation of this bimodal $G(\Gamma)$ was performed by assuming: (1) that both peaks were caused by translational motion and (2) that the low-frequency peak alone was translational in origin. If a plot similar to Figures 3 and 4 is made, the latter assumption results in a value of $b = 0.64$. Therefore, we question the source of the high-frequency component of $G(\Gamma)$ as translational motion.

The agreement between the value of $b = 0.64$, deduced for PAM-A and PAM-B, with the value of b' measured by Francois et al. (6) is reassuring. The determination of k_T at two different molecular weights by the transformation process also yields similar values at higher scattering angles, 5.2×10^{-4} and 4.5×10^{-4} $(cm^2/sec)(mole/g)^{-b}$ for PAM-B and PAM-A in H_2O, respectively.

Finally, we may compare our results for $f(M)$ with those from size exclusion chromatography (SEC). SEC results for PAM-A in salt solution indicate that $M_n : M_w : M_z \simeq 1 : 2 : 4$, whereas light scattering studies show that M_z / M_w is between 1.6 and 2.5, depending on concentration, angle, and sample. Comparison of the $f(M)$ curves directly shows striking differences caused by the biasing of light scattering toward the larger species and SEC toward the lower molecular weight components (19); Figure 5 shows this comparison. Apparently, light scattering deduced $f(M)$ does not include any molecules with $M \leq 6.9 \times 10^4$ g mole^{-1}. Both curves show a tail toward higher molecular weights, but the area under these tails is very much smaller in the SEC case.

6. SUMMARY OF INVERSION TECHNIQUES

In particle size analysis, we concentrate on a determination of $G(\Gamma)$ using approximations that permit us to make a meaningful Laplace inversion. The size distributions are then expressed in terms of equivalent Stokes radius ($r_h = k_B T / 6\pi \eta_s D$, with η_s being the solvent viscosity). In the size transform, interparticle interactions and particle shape are usually neglected. The previous sections emphasized how we should take into account those effects, especially for polydisperse macromolecular solutions. It should be noted that the particle scattering factor contributes appreciably whenever $K r_g$ approaches 1. Nevertheless, the determination of $G(\Gamma)$ remains a central issue in such a technique.

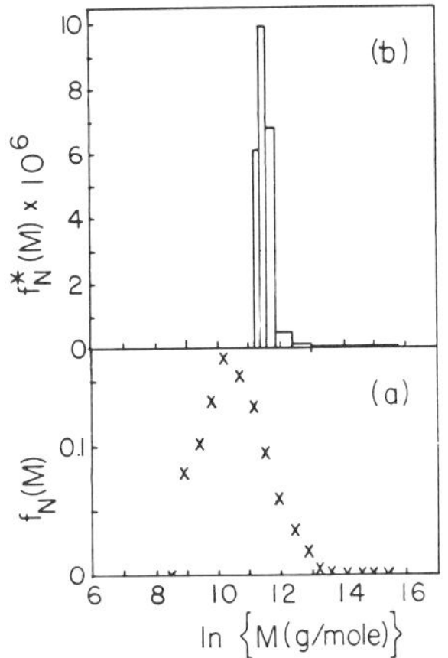

Figure 5. (*a*) The number fraction of PAM-A in 0.5 *N* NaNO$_3$, $C = 0.4$ wt %. The molecular weight distribution is obtained from SEC measurements. The distribution was converted from a weight distribution, as measured by the refractive index used to monitor the SEC column. The peak heights were normalized to unity. The upper histogram represents the transformation presented previously (19) [the open cirlce at $\theta = 120°$ with $\sin^2(\theta/2) = 0.75$ in Figure 4]. Recall that a high-frequency component complicated this histogram. In this transformation, we use only te low-frequency portion of $G(\Gamma)$. The two $f_w(M)$'s are apparently quite different, yet parameters we compute from them are in agreement. Superscript * denotes unnormalized distribution function.

We now have a variety of methods of data analysis for approximating $G(\Gamma)$ in Equation 1. These include the cumulants method (1), the histogram method (2, 3), the Pearson method (20, 21), the splines method (22), the Laplace inversion method (23, 24) and the multiexponential method (25–28). The cumulants method can provide information on the average linewidth and the variance, whereas higher-order moments are usually not trustworthy. The Pearson method is limited to unimodal distributions and involves extensive numerical computations. The histogram method represents the zeroth order splines method. There are similarities between the histogram method, the splines method, and the multiexponential method, especially when exponential sampling is used. Mathematical justifications of the Laplace inversion problem have been discussed by McWhirter and Pike (25) and by Provencher (23). Basically, with experimental data measured over finite time ranges, the solution becomes eigenfunctions of the Laplace transform. The ill-posed problem can be optimally stabilized using a constrained regularization method that finds the smoothest *non-negative* solution consistent with the data. In most cases, such as those encountered in molecular polydispersity problems, all of the methods work sufficiently well to provide appropriate estimates of $G(\Gamma)$. The more practical solution is often related to the simplicity of the algorithm used in the computer program and the speed with which estimates of $G(\Gamma)$ can be achieved. Therefore, the trend is toward linear programming using the singular value decomposition technique (29). The improvements will permit on-line determinations of polymer molecular weight distributions of known polymer/solvent systems using photon correlation and low-cost microprocessors.

NOTATION

A_2	Second virial coefficient, mole cm^3 g^{-2}.
b	Exponent relating molecular weight to translational diffusion coefficient, D_0, obtained from dynamic light scattering measurements.
b'	Exponent relating molecular weight to radius of gyration, r_g, obtained from static light scattering measurements.
C	Total concentration, $\sum_i C_i$.
C_i	Concentration of M_i.
D	Translational diffusion coefficient, cm^2 sec^{-1}.
D_0	Diffusion coefficient at infinite dilution, cm^2 sec^{-1}.
E	Electric field vector.
E^*	Complex conjugate of electric field vector.
$f(M)$	Weight molecular weight distribution.
$f_N(M)$	Number molecular weight distribution.
$g^{(1)}(\tau)$	First-order electric field correlation function.
$G(D)$	Diffusion coefficient distribution.
$G(\Gamma)$	Linewidth distribution.
$G'(\Gamma)$	Normalized linewidth distribution.
$\langle I(\theta) \rangle$	Time averaged total scattered intensity at scattering angle, θ.
k_D	Diffusion second virial coefficient, cm^3 g^{-1}.
k_T	Coefficient relating molecular weight to translational diffusion coefficient, D_0, obtained from dynamic light scattering measurements, cm^2 sec^{-1} g^{+b} mole^{-b}.
k_T'	Coefficient relating molecular weight to radius of gyration, r_g, obtained from static light scattering measurements, cm mole$^{b'}$ g$^{-b'}$.
K	Magnitude of momentum transfer vector, $4\pi n_0 \sin(\theta/2)/\lambda_0$.
K', K''	Constants.
M_i	Molecular weight of component i.
ΔM_i	Molecular weight increment.
$\overline{M}_i$	Center of histogram step in molecular weight space.
M_n	Number average molecular weight.
M_w	Weight average molecular weight.
M_z	Z-average molecular weight.
n_0	Refractive index of solvent.

$P(\theta, M_i)$ Particle scattering factor.

r_g Radius of gyration, Å.

$\langle r_g^2 \rangle_{ls}^{1/2}$ Root mean-square $P(\theta, M)$ weighted z-average radius of gyration, Å.

$\langle r_g^2 \rangle_z^{1/2}$ Root mean-square concentration independent z-average radius of gyration, Å.

$\langle r_g^2(C) \rangle_z^{1/2}$ Root mean-square concentration dependent radius of gyration, Å.

r_h Stokes radius, Å.

η_s Solvent viscosity.

λ_0 Wavelength of incident light in vacuum.

θ Scattering angle.

τ Delay time, sec.

Γ Linewidth.

$\bar{\Gamma}_i$ Center of histogram in Γ-space.

$\Delta\Gamma_i$ Step width in Γ-space.

REFERENCES

1. D. E. Koppel, *J. Chem. Phys.*, **57**, 4814 (1972).

2. B. Chu, Esin Gulari, and Erdogan Gulari, *Phys. Scr.*, **19**, 476 (1979).

3. Esin Gulari, Erdogan Gulari, Y. Tsunashima, and B. Chu, *J. Chem. Phys.*, **70**, 3965 (1979).

4. Erdogan Gulari, Esin Gulari, Y. Tsunashima, and B. Chu, *Polymer*, **20**, 347 (1979).

5. B. Chu and Esin Gulari, *Macromolecules*, **12**, 445 (1979).

6. J. Francois, D. Sarazin, T. Schwartz, and G. Weill, *Polymer*, **20**, 969 (1979).

7. J. Klein and K.-D. Conrad, *Makromol. Chem.*, **181**, 227 (1980).

8. B. Berne and R. Pecora, *Dynamic Light Scattering*, Wiley-Interscience, New York, 1976.

9. J. McWhirter, *Opt. Acta*, **27**, 83 (1980).

10. T. A. King, A. Knox, W. I. Lee, and J. D. G. McAdam, *Polymer*, **14**, 151 (1973).

11. H. Yamakawa, *Modern Theory of Polymer Solutions*, Harper and Row, New York, 1971.

12. D. Caroline, *Polymer*, **22**, 576 (1981).

13. J. Raczek and G. Meyerhoff, *Macromolecules*, **13**, 1251 (1980).

14. W. Burchard, *Polymer*, **20**, 577 (1979).

15. C. C. Han and F. L. McCrackin, *Polymer,* **20,** 427 (1979).

16. M. B. Huglin (Ed.), *Light Scattering from Polymer Solutions,* Academic Press, New York and London, 1972.

17. A. DiNapoli, Ph.D. Dissertation, Stony Brook (1981).

18. B. Chu and A. DiNapoli, in Barton Dahneke (Ed.) "Measurement of Suspended Particles by Quasielastic Light Scattering", John Wiley & Sons, New York, 1983, pp. 81–105.

19. A. DiNapoli, B. Chu, and C. Cha, *Macromolecules,* **15,** 1174 (1983).

20. K. Pearson, *Phil. Trans. R. Soc.,* **185,** 71 (1894).

21. W. P. Elderton, *Frequency Curves and Correlation,* Harren Press, Washington, D.C., 1953, Chapter 2.

22. G. B. Stock, *Biophys. J.,* **16,** 535 (1976); **22,** 79 (1978).

23. S. W. Provencher, *J. Chem. Phys.,* **64,** 2772 (1976); *Makromol. Chem.,* **180,** 201 (1979).

24. S. W. Provencher, J. Hendrix, L. De Maeyer, and N. Paulussen, *J. Chem. Phys.,* **69,** 4273 (1978).

25. J. McWhirter and E. R. Pike, *J. Phys. A,* **11,** 1729 (1978).

26. E. R. Pike, in S. H. Chen, B. Chu, and R. Nossal (Eds.), *Scattering Techniques Applied to Supramolecular and Nonequilibrium Systems,* Plenum Press, New York, 1981, pp. 179–200.

27. D. Sornette and N. Ostrowsky, in S. H. Chen, B. Chu, and R. Nossal (Eds.), *Scattering Techniques Applied to Supramolecular and Nonequilibrium Systems,* Plenum Press, New York, 1981, pp. 755–758.

28. N. Ostrowsky, P. Parker, E. R. Pike, and D. Sornette, *Opt. Acta,* **28,** 1059 (1981).

29. C. L. Lawson and R. J. Hanson, *Solving Least Squares Problems,* Prentice-Hall, New Jersey, 1974.

CHAPTER

5

PARTICLE AND DROPLET SIZING
USING FRAUNHOFER DIFFRACTION

BRUCE B. WEINER

Brookhaven Instruments Corporation
Ronkonkoma, New York

1. INTRODUCTION

Particle sizing is important in a large number of practical applications, and reasons for this circumstance are not hard to find. With few exceptions, raw materials are rarely found in nature in a form suitable for end use. Typically, such raw materials are converted into finished products by chemical and physical processes. Such processes include grinding, dissolving, evaporating, heating, and chemically reacting the raw material. Particle size is a property that significantly affects these processes.

Raw materials are often pulverized for convenience in handling. Here, particle size affects mechanical properties such as packing and flowing. Some materials find their end use as specifically tailored size distributions of particles because size, among other properties, determines usefulness. With few exceptions, almost every solid material either starts out as a powder, ends up as a powder, or goes through at least one powder stage during manufacture.

With so many substances of practical interest exhibiting the effects of particle size, it is no wonder that the number of techniques and instruments (1–3) for particle sizing abound. Just about every well-known physical principle has been applied in one way or another (see Chapters 1 and 2).

Techniques involving optical principles have certain advantages for particle sizing. They are nonintrusive, are much faster than mechanical techniques, do not require conducting media, and, aside from possible effects in sample handling, do not subject the particle to large shearing forces. A large number of optical techniques are available. Chabay and Bright (4) reference 32 instruments based on 15 different optical measurements, just for sizing aerosol particles.

Except for a very few simple shapes, analytical expressions relating size to a physically measurable quantity do not exist. If shape information is required, then an imaging technique is necessary.

Imaging techniques include the traditional forms of microscopy and photography, as well as the more recent video and holographic techniques. The advantages of good, representative images are many: size, shape, number of particles, and, possibly, composition information.

There are also many disadvantages to image analysis. Good, representative images may involve extensive sample preparation. In some cases, the preparation involved (e.g., drying) may destroy or alter the original integrity of the particle. Optical depth of field adds to the normal sampling problems. The analysis of images is not as straightforward as it may seem. To obtain reliable statistics, one must count a sufficient number of particles, at least many hundreds (5). Manual analysis is not only time consuming, but also subject to operator fatigue and bias. Automatic image analysis is better, but it negates one of the main advantages of the simple microscopic technique—cost.

Even with good, representative images and an automatic image analyzer, one has to choose from a large number of totally empirical schemes for shape analysis (6). Every conceivable graphical and geometric method has been applied. Every length on the image has been used as a statistical descriptor of size: longest cord, distance between two parallel lines, average cord length, and so on. Every ratio of two lengths has been named (elongation, flakiness, thickness, bulkiness, surface factor, etc.) and statistically analyzed for relevance. Although this type of shape analysis is important for some types of particle size analysis, it is not necessary for many others.

Because of the difficulties involved in characterizing particle shape, all automatic particle sizers, with the exception of an automated image analyzer, assume an equivalent sphere diameter. This concept is discussed further in Section 7.4.

For particles larger than 1 or 2 μm, the optical principle of diffraction, specifically Fraunhofer diffraction (FD), can be utilized (7, 8). It is fast, needs no calibration, and is based on well-known, rigorous, optical principles. When configured properly, an instrument of this type can be used for both solid particles and liquid droplets, either of which may be suspended or flowing in either a gas or liquid. This generality has been exploited in one of the commercially available instruments manufactured by Malvern Instruments, details of which are described in this chapter.

2. THEORY

To understand the advantages and disadvantages of FD for particle sizing, it is useful to review some general aspects of light scattering.

2.1. Definitions

When a beam of incident light interacts with a particle, it can be absorbed, scattered, or transmitted. Conservation of energy demands that the sum of the energy in the three interactions equal the incident energy. Defining the amount of extinction of the beam as the difference between the incident and transmitted energy, it follows that:

$$C_{\text{ext}} = C_{\text{sca}} + C_{\text{abs}} \tag{1}$$

where C is the integrated or total cross section for extinction, scattering (includes reflection, refraction, diffraction), and absorption, respectively. For example, the cross section C_{sca} represents the total energy/unit time or power scattered from the main beam, integrated over all directions. The units of C are length squared. The cross sections represent effective cross-sectional areas of the particle.

It is convenient to define dimensionless ratios of these effective cross sections to the cross-sectional area of the particle. For a sphere of radius r,

$$Q = \frac{C}{\pi r^2} \tag{2}$$

where Q is the cross-sectional efficiency. Whenever FD is applicable, the scattering cross-section efficiency Q_{sca} has a particularly simple value: it is always equal to 2.

The transmittance T is defined as the ratio of the intensity of the transmitted beam I_T to the intensity of the incident beam I_0. Obviously, T is a function of the total extinction cross section. The relationship is given by the Beer–Lambert law:

$$T = \frac{I_T}{I_0} = \exp(-NC_{ext}l) \tag{3}$$

where N is the number of particles per unit volume, and l is the path length.

Since C_{ext} is the sum of two terms, Equation 3 factors into a product of two terms: one for scattering and one for absorption. If no absorption takes place, then extinction can only be due to scattering. It is usual to define the turbidity, τ, as

$$\tau = NC_{sca} \tag{4}$$

In this case, the transmittance is only caused by scattering and can be written as

$$T = \exp(-\tau l) \tag{5}$$

2.2. General Case

The theory for calculating the extinction cross section from first principles is called Lorenz–Mie theory (9). It starts from Maxwell's electromagnetic equations and proceeds to exact answers for the cross-section efficiencies in terms of a size parameter, x, the index of refraction of the particle relative to the medium in which it is suspended, m, and the angle of observation, θ. The index of refraction can be a complex number, the imaginary part of which gives rise to absorption.

Through the size parameter, measurements of the scattered light intensity or turbidity yield information on particle size. The size parameter for a sphere of diameter d is

$$x = \frac{2\pi r}{\lambda} = \frac{\pi d}{\lambda} \tag{6}$$

where λ is the wavelength of the light in the suspending medium.

2.3. Limiting Cases

Although the Lorenz–Mie theory is exact, it does not, in general, lead to simple, analytical solutions relating particle size to optical measurements. However, there are well-known limiting cases—Rayleigh scattering and diffraction—which are much simpler. Sometimes the criteria for the applicability of these limiting cases are stated simply in terms of diameter and wavelength as

$$
\begin{aligned}
d \ll \lambda \qquad &\text{Rayleigh scattering} \\
d \approx \lambda \qquad &\text{Lorenz–Mie theory} \\
d \gg \lambda \qquad &\text{Diffraction theory}
\end{aligned}
\tag{7}
$$

and sometimes a rule of thumb is given, such as

$$
\begin{aligned}
d < \frac{\lambda}{10} \qquad &\text{Rayleigh} \\
d > 4\lambda \qquad &\text{Diffraction}
\end{aligned}
\tag{8}
$$

A more useful parameter for discerning limiting cases takes into account both the size parameter and the relative index of refraction. Such a parameter is suggested by Van de Hulst (10):

$$p = \frac{2\pi d \, |\, m - 1 \,|}{\lambda} \tag{9}$$

where

$$
\begin{aligned}
p < 0.3 \qquad &\text{Rayleigh} \\
p \approx 1 \qquad &\text{Lorenz–Mie} \\
p \gg 30 \qquad &\text{Fraunhofer}
\end{aligned}
\tag{10}
$$

However, the criteria do not represent exact transitions. They can only suggest when limiting cases may be applicable. In regions of overlap, one can calculate the exact answer from the Lorenz–Mie theory and compare it to the results from the simpler, limiting cases. Jones (11) has in fact done this for spheres.

2.4. Fraunhofer as a Limiting Case

Jones restricted his calculations to comparing the difference between exact theory and the FD result for forward scattering. Since the instrument to be discussed is limited to forward scattering, his results are directly applicable.

Jones expressed his results in terms of error contour charts, as a function of the real part of the relative index of refraction and the size parameter. The error represents the percentage difference between the exact size parameter and the one calculated from the Fraunhofer predicted diameter at an angle, calculated from the exact theory, at which the intensity of the forward scattered lobe fell to $\frac{1}{2}$ its zero-angle value. He calculated results for two cases: nonabsorbing and absorbing (the imaginary part of the relative index equal to 0.1).

The contour charts show that in the region of overlap between Lorenz–Mie and FD theory, the situation is complicated. There are "islands" of error of 20% or greater in the region bounded by $m \leq 2$ and $x \geq 20$. The latter corresponds to a diameter of 4 μm in air when $\lambda = 0.6328$ μm. However, surrounding some of these error "islands" are regions of 10% and zero error.

For $x \geq 20$ ($d \geq 4$ μm in air) and $m \geq 1.3$, the errors are less than 20%. Errors greater than 20% can also occur for larger particles if $m \leq 1.2$.

The contour charts for the case with absorption are less complicated. Many of the large error "islands" disappear. Errors of 20% do not occur until $x \leq 8$ ($d \leq 1.6$ μm in air) and $m \leq 1.5$. For m between 1.2 and 1.5, errors of less than 10% can occur for x as small as 4 (d as small as 0.8 μm in air).

The overall result from Jones' work is this: FD theory can be applied to particles as small as 1 or 2 μm in some cases and, in most cases, to 5-μm particles with errors always less than 20%.

2.5. Classical Fraunhofer Diffraction Theory

Fraunhofer diffraction theory is derived from fundamental optical principles that are not concerned with scattering (12). Sometimes the theory

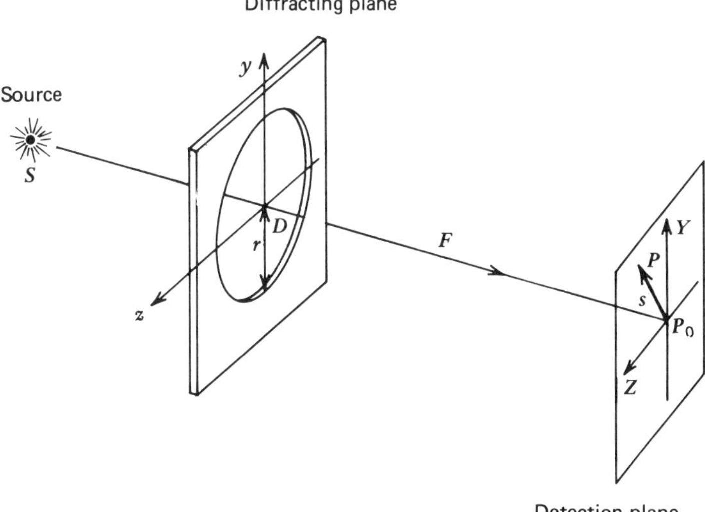

Figure 1. Schematic showing the source of radiation impinging on a diffracting plane containing an aperture (or opaque disk or cross section of sphere) with radius r and detection plane. For Fraunhofer diffraction, the distance SD times the wavelength must be large compared to the area of the scattering object. Likewise, the distance F times the wavelength must also be large compared to the object's scattering area. The distance from the center of the detector to a point P in the same plane is designated s.

is described as accounting for the "bending" of light caused by the edges of an aperture. In fact, the basic Fresnel–Kirchhoff theory of diffraction, one limit of which is FD, results in an integration of the incident electric field over the entire aperture. Babinet's theorem is invoked to obtain the diffraction pattern of an opaque object whose cross-sectional area and shape are equal to that of the aperture. The diffraction pattern (in the Fraunhofer limit) for a complementary, opaque object is the same, except for the geometric shadow cast by the object.

Two requirements are needed to obtain the Fraunhofer limit of the Fresnel–Kirchhoff diffraction theory. First, the area of the aperture or object must be much smaller than the wavelength of light times the distance from the source point to the diffracting plane (Figure 1). Second, the area must also be smaller than the wavelength times the distance from the observation plane to the diffracting plane. For these reasons, FD is known as far-field diffraction.

If a plane wave is incident on the diffracting plane, the source point can be considered at infinity, and the first requirement is satisfied. The second requirement can be achieved by placing the detection plane far away, or by using a lens to focus the diffracted light onto a detector in the focal plane of the lens.

Since the objects of interest here are all rather small particles, the use of the lens is not primarily to satisfy the distance requirement. Because the detector is in the focal plane of the lens, the diffraction pattern of a moving particle will be stationary. Thus, particle sizing will not be biased by the speed of the particle. (A sampling bias resulting from differential particle speeds may be significant and is discussed in Section 7.3; however, the diffraction pattern for any single particle is independent of particle velocity when the focal plane is used for detection.)

2.6. Fraunhofer Diffraction Theory for Distributions of Spheres

The well-known result for the intensity of the diffraction pattern for an aperture, disk, or sphere of radius r, is

$$I = I_0 \left[\frac{2J_1(x)}{x} \right]^2 \tag{11}$$

where I_0 is the intensity at the center of the pattern, J_1 is the first-order spherical Bessel function, and x is given by

$$x = \frac{2\pi rs}{\lambda F} \tag{12}$$

where s is the radial distance in the detection plane as measured from the optical axis and F is the focal length of the lens (Figure 1).

Equation 11 is plotted in Figure 2. The function is known as the Airy function, after the British astronomer who first derived it. The pattern consists of a series of bright and dark concentric rings surrounding the bright central disk whose radius is $x = 3.83$.

Measuring the intensity pattern for the purpose of inferring a size distribution is difficult (13); measuring and analyzing the light energy distributed over a finite area of the detector is more fruitful (8). For this purpose, Equation 11 is integrated to obtain the fraction of the light energy contained within a circle of radius s on the detector (radial distance in detection plane). The result is

$$L = 1 - J_0^2(x) - J_1^2(x) \tag{13}$$

where J_0 is the zero-order spherical Bessel function.

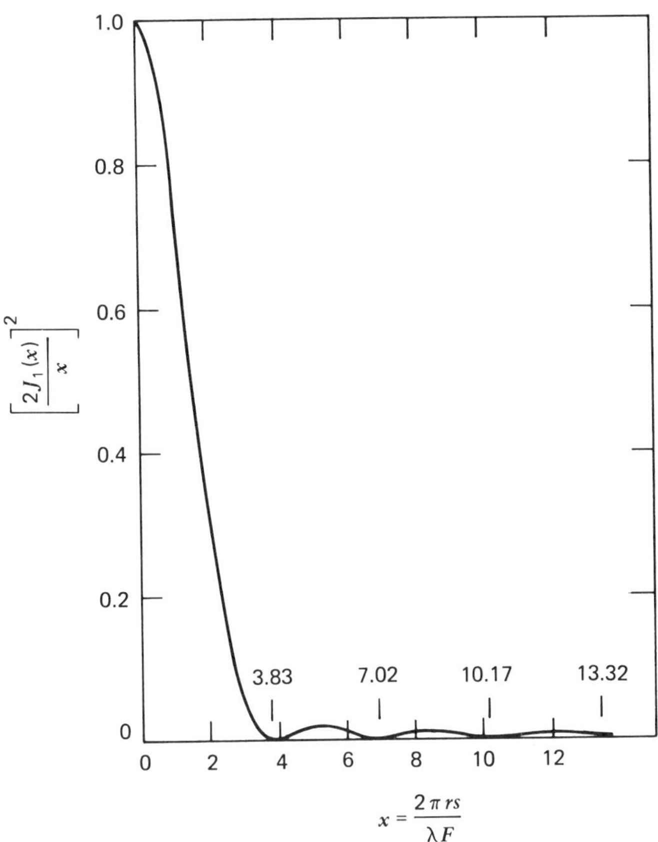

$$x = \frac{2\pi rs}{\lambda F}$$

Figure 2. Fraunhofer diffraction pattern for circular aperture or opaque disk (equation 11) where r is the particle radius, s is the radial distance in the detection plane, and F is the focal length of the lens.

For the instrument described later, the detector consists of a series of rings (Figure 3). It is, therefore, more appropriate to ask for the light energy falling between two radii, s_1 and s_2, owing to a single particle of radius r. This is

$$L_{s_1,s_2} = C\pi r^2 [J_0^2(x) + J_1^2(x)]_{s_1} - [(J_0^2[(x) + J_1^2(x)]_{s_2} \qquad (14)$$

where πr^2, the particle cross-sectional area, is proportional to the light energy falling on the particle and C is an optical constant that depends on the power of the light source and on the detector sensitivity.

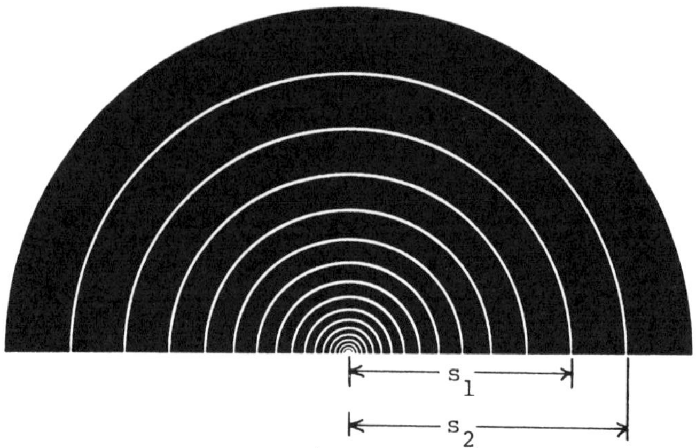

Figure 3. Schematic of detector that consists of photosensitive rings (dark) separated by equal thickness insulating gaps (light). Distances in the detector plane are measured from the central optical axis—a hole in the detector plane—behind which is a separate photodiode.

For N particles, each of radius r, the appropriate expression, neglecting multiple diffraction, is N times Equation 14. Therefore, the light energy falling between s_1 and s_2, owing to a large number of particles of different sizes, is

$$L_{s_1,s_2} = C\pi \sum_{i=1}^{M} N_i r_i^2 \{[J_0^2(x_i) + J_1^2(x_i)]_{s_1} - [J_0^2(x_i) + J_1^2(x_i)]_{s_2}\} \quad (15)$$

where the size distribution has been divided into M size classes.

At this point, it appears that the formalism allows a conversion from the measured L-values to the number distribution represented by the set of N_i. For reasons discussed in Section 6.4, this is not a useful approach.

It is more convenient to rewrite Equation 15 in terms of the mass or weight of particles W. This is accomplished by assuming that the density, ρ, of a particle is independent of size. Namely,

$$N_i = \frac{3W_i}{4\pi r^3 \rho} \quad (16)$$

Equation 15 becomes,

$$L_{s_1,s_2} = K \sum_{i=1}^{M} \left(\frac{W_i}{r_i}\right) \{[J_0^2(x_i) + J_1^2(x_i)]_{s_1} - [J_0^2(x_i) + J_1^2(x_i)]_{s_2}\} \quad (17)$$

where K contains the optical constant C as well as the density.

This is the basic relationship between measured light energy and size distribution. The instrument described in Section 3 yields 15 L-values, from which 15 W-values are calculated corresponding to a weight distribution over 15 size classes. The limits of each size class depend on the ring radii s_i, which have been chosen to take maximum advantage of the dynamic range of measured signals. Although Equation 11 and Figure 2 show that all particle sizes diffract light intensity into, essentially, all rings, the light energy distribution, Equation 13, has a maximum at

$$x = \frac{2\pi rs}{\lambda F} = 1.375 \tag{18}$$

The size class limits are calculated to correspond to this equation, given the set of ring radii. Thus, Equation 17 is one of a set of 15 equations. The 15 size classes correspond to 15 sets of s_1 and s_2. The largest size classes are usually quite broad, and using the midpoint of the size class to represent r_i can lead to problems in resolution. For this reason, each of the 15 terms in Equation 17 is further subdivided into a large number of equally spaced, subsize classes, and the midpoint of each of these small size ranges yields an r_i. It is assumed that the distribution by weight in any full class size is constant.

Given the above formalism and assumptions, it is natural to try a matrix inversion to solve for the set of values from Equation 17. However, the dynamic range of the 225-element matrix and the experimental noise in the 15 L-values often leads to nonphysical results.

Another approach is to use the least-squares criteria. Here, an initial set of 15 W_i's is either estimated from the raw data (model independent) or calculated from an assumed functional form of the size distribution (model dependent). These 15 values are used to calculate a set of 15 L's. The sum of the squares of the differences between the 15 measured and calculated L-values is the error corresponding to the initial set. Various algorithms are used to obtain, iteratively, sets of W's, until the least-square error has been minimized.

3. INSTRUMENT CONFIGURATION

Figure 4 shows the basic optical setup. A 2-mW He–Ne, unpolarized laser beam is spatially filtered, expanded to 9 mm, and collimated. Particles are allowed to move across this beam. Diffracted and transmitted light are focused by a lens onto the detector, which is in the focal plane of the lens. The smallest particles that can be measured, on the order of 1 μm,

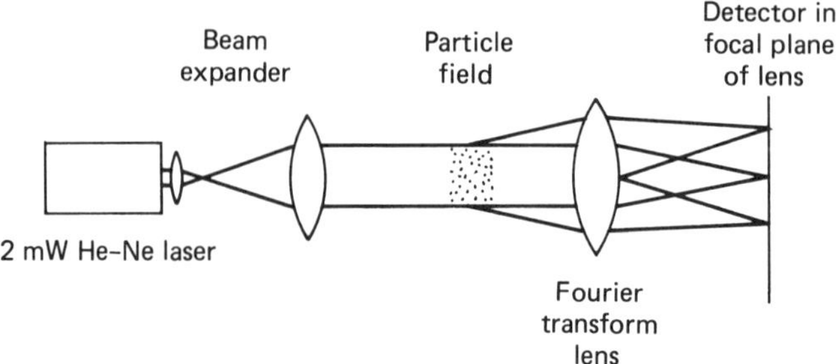

Figure 4. Optical schematic of Fraunhofer diffraction instrument used for droplet and particle size analysis.

diffract light to the largest angle. Since this angle is only 11°, the technique is similar to forward angle scattering.

Figure 5 is a photograph of the exterior of a system used for airborne droplet and particle size measurements. Figure 6 depicts an instrument used for measuring liquid-borne droplets and particles. A variable speed pump, variable speed stirrer, and ultrasonic mixing tank allow for a variety of sample handling techniques. Inlet and outlet ports are standard features which are useful for on-line measurements when process control is necessary. Otherwise, sample and carrier liquid are added manually to the mixing tank.

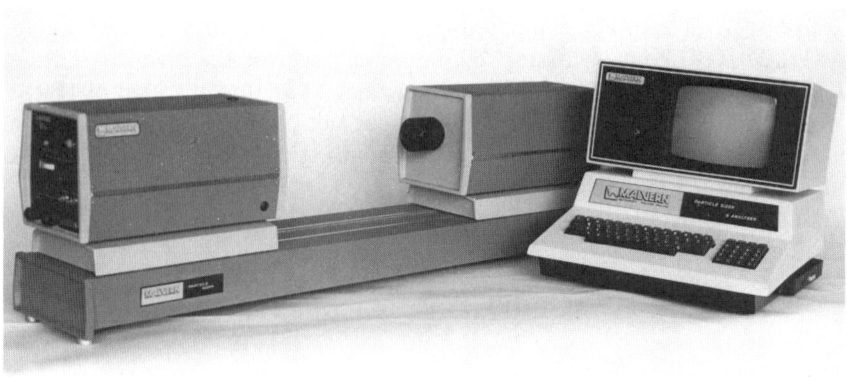

Figure 5. Malvern laser diffraction instrument used for airborne particle size distribution measurement. Photograph courtesy of Malvern Instruments Ltd.

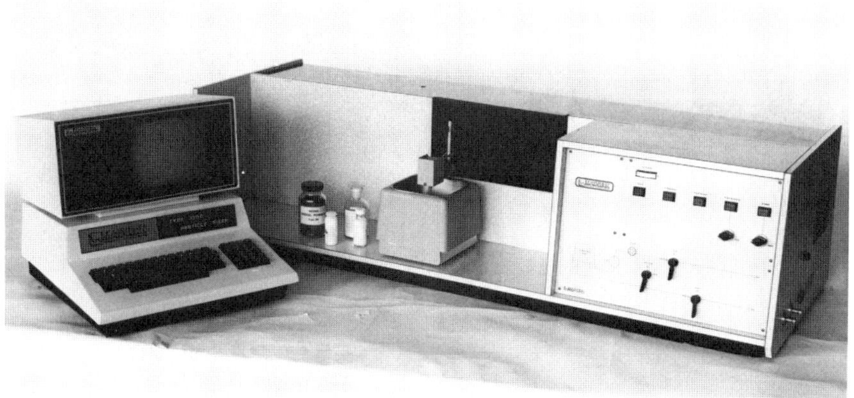

Figure 6. Malvern laser diffraction instrument used for liquid-borne particle size distribution measurement. Photograph courtesy of Malvern Instruments Ltd.

Three focal length lenses are included: 63, 100, and 300 mm. The lens focuses the light onto a 30-element, concentric, light sensitive ring detector with a hole in the center. Behind the hole is a photodiode used for alignment and for measuring I_T/I_0 (see Figure 3). The radii of the rings are such that the size ranges covered are 1.2–118 µm, 1.9–188 µm, and 5.8–564 µm, respectively. Three additional lenses are available which extend the range to over 1800 µm. The size range covered by any one lens is divided into 15, logarithmically increasing size classes with the ratio of the largest to smallest diameter about 100:1. Table 1 shows the size classes for each focal length. The user is advised to select a lens that covers the size range and resolves its most prominent features adequately.

An experiment consists of taking data for background and then for signal plus background. Data are taken by sweeping all 30 rings a number of times. During a single sweep, ring signals are collected serially, digitized, and stored in the computer for averaging with the next sweep. The total number of sweeps is user-selectable and is chosen to match the total time the sample will be present in the beam. A minimum of 10 sweeps is recommended to smooth out statistical fluctuations. Typically, a few hundred sweeps are taken to ensure that a representative, randomly oriented sample from all size classes has been measured.

At the end of the experiment, the average background is subtracted from the average signal plus background, and each successive pair of ring data is averaged. These 15 data points are then normalized. They now represent the 15 measured L-values in Equation 17.

Table 1. Size Classes (μm) Versus Focal Length (mm) for Malvern Fraunhofer Diffraction Instruments

63 mm Focal Length		100 mm Focal Length		300 mm Focal Length	
Upper	Lower	Upper	Lower	Upper	Lower
118.4	54.9	188.0	87.2	564.0	261.6
54.9	33.7	87.2	53.5	261.6	160.4
33.7	23.7	53.5	37.6	160.4	112.8
23.7	17.7	37.6	28.1	112.8	84.3
17.7	13.6	28.1	21.5	84.3	64.6
13.6	10.5	21.5	16.7	64.6	50.2
10.5	8.2	16.7	13.0	50.2	39.0
8.2	6.4	13.0	10.1	39.0	30.3
6.4	5.0	10.1	7.9	30.3	23.7
5.0	3.9	7.9	6.2	23.7	18.5
3.9	3.0	6.2	4.8	18.5	14.5
3.0	2.4	4.8	3.8	14.5	11.4
2.4	1.9	3.8	3.0	11.4	9.1
1.9	1.5	3.0	2.4	9.1	7.2
1.5	1.2	2.4	1.9	7.2	5.8

600 mm Focal Length		800 mm Focal Length		1000 mm Focal Length	
Upper	Lower	Upper	Lower	Upper	Lower
1128.0	523.2	1503.9	697.6	1879.9	872.0
523.2	320.7	697.6	427.6	872.0	534.5
320.7	225.6	427.6	300.8	534.5	376.0
225.6	168.6	300.8	224.8	376.0	281.0
168.6	129.3	224.8	172.4	281.0	215.5
129.3	100.3	172.4	133.8	215.5	167.2
100.3	78.0	133.8	104.0	167.2	130.0
78.0	60.7	104.0	80.9	130.0	101.1
60.7	47.3	80.9	63.1	101.1	78.8
47.3	36.9	63.1	49.2	78.8	61.5
36.9	29.0	49.2	38.6	61.5	48.3
29.0	22.8	38.6	30.4	48.3	38.0
22.8	18.1	30.4	24.1	38.0	30.2
18.1	14.5	24.1	19.3	30.2	24.1
14.5	11.6	19.3	15.5	24.1	19.4

Sample concentration = 0.0064 % by volume
Obscuration = 0.25

Size Classes	Size Band		Cumulative Weight Below	Weight in Band	Cumulative Weight Above	Light Energy	
	Upper	Lower				Computed	Measured
15	188.0	87.2	100.0	0.0	0.0	196	266
14	87.2	53.5	95.0	5.0	0.0	274	374
13	53.5	37.6	58.0	37.1	5.0	359	489
12	37.6	28.1	58.0	0.0	42.0	419	581
11	28.1	21.5	58.0	0.0	42.0	470	609
10	21.5	16.7	58.0	0.0	42.0	555	706
9	16.7	13.0	58.0	0.0	42.0	805	988
8	13.0	10.1	15.1	42.8	42.0	1160	1464
7	10.1	7.9	13.6	1.5	84.9	1398	1787
6	7.9	6.2	13.6	0.0	86.4	1430	1927
5	6.2	4.8	13.6	0.0	86.4	1146	1576
4	4.8	3.8	13.6	0.0	86.4	1027	1105
3	3.8	3.0	13.6	0.0	86.4	1514	1262
2	3.0	2.4	1.5	12.1	86.4	1929	1990
1	2.4	1.9	0.0	1.5	98.5	2047	2047

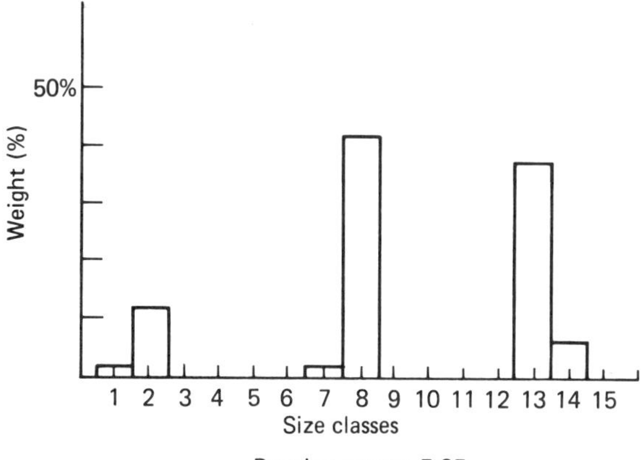

Figure 7. Tabular and graphical output of laser diffraction instrument for particle sizing showing trimodal distribution of standard latices. Used with permission from reference 14.

The software allows for a variety of user-selectable features. The data may be analyzed in terms of a histogram with 15 size classes. Figure 7 illustrates this for a trimodal distribution of standard latices (14). Tabular data include the cumulative percent by weight below the lower size class limit, the percent by weight in the size class, and the cumulative percent by weight above the upper size class limit. If the path length has been input, then the obscuration $(1 - T)$ is used to calculate the sample concentration in terms of the total volume of particles as a percentage of the beam volume. The log error is the logarithm of the sum of the squares of the differences between the computed and measured L-values. It is this error that is minimized in the least-squares search.

Data sets are stored in any one of 32 RAM memory blocks. Mass storage of data is available on magnetic tape or, optionally, on diskettes. Run numbers and time of day are part of each data set. User comments can be appended and printed. Graphs of the three types of tabular data, one of which is shown in Figure 7, may be printed.

The program is menu-driven with a summary of the most often used commands available for display on the CRT. The user can choose to display the data in real time as the particles pass through the beam. During the least-squares search, the successive iterations are graphically displayed, and the user can also display the graphs of the three types of data output. Any of the graphs on the CRT can be printed.

The command structure allows for looping. This is most effective when sequential sets of data are to be taken automatically and stored in successive data block memories. Looping is also useful for automatic data analysis and printing and for automatically storing data on and recalling from magnetic tape or diskette.

With live display of data, it takes 35 msec to serially sweep all 30 rings. For faster, snapshot-like measurements, a new system is available that takes data from all 30 rings, in parallel, in 10 μsec. This may be repeated at 35-msec intervals. This extra speed will be utilized in high-speed, pulsed-spray applications.

If the data are unimodal, and if the user wishes to represent them by an assumed functional form for the size distribution by weight, then the computer allows the user to choose from one of three, two-parameter models. These are the log-normal,

$$dW = \frac{1}{\sqrt{2\pi}\,\ln(\sigma)} \left\{ \exp\left[-\left(\frac{\ln(x/\bar{x})}{2\ln(\sigma)} \right)^2 \right] \right\} d\ln(x) \tag{19}$$

the normal,

$$dW = \frac{1}{\sqrt{2\pi}\,\sigma} \left\{ \exp\left[-\frac{(x - \bar{x})^2}{2\sigma^2} \right] \right\} dx \tag{20}$$

and the Rosin–Rammler,

$$dW = \sigma \left(\frac{x^{\sigma-1}}{\bar{x}^\sigma} \right) \left\{ \exp\left[-\left(\frac{x}{\bar{x}} \right)^\sigma \right] \right\} dx \tag{21}$$

In each case, $\bar{x}$ and σ represent the two parameters in the model and dW is the weight fraction between diameters x and $x + dx$. In the normal equation, they are defined as the mean and standard deviation of the weight distribution, respectively. In the log-normal equation, $\bar{x}$ is the geometric mean diameter by weight, which is equivalent to the median diameter of the weight distribution, and σ is the geometric standard deviation. The log-normal equation is almost universally acceptable as a model for sizing particles, except in the field of fuel sprays, where there are a number of models (15). The Rosin–Rammler is the simplest model, and it agrees fairly well with experimental data (16). The $\bar{x}$ and σ in this equation can be related to such standard statistics as the average, median, and standard deviation through the gamma function, which is obtained upon integration of the Rosin–Rammler function when moments are calculated.

If one of the three models is chosen to represent the data, the user can instruct the computer to calculate 32 equally spaced size classes using the model parameters. It is also possible to modify the user-selectable spacings such that data are presented in powers of $2^{1/2}$ or $2^{1/4}$ for comparison with standard sieve-size data. Tabular and graphical data, with these user-selectable size classes, can be displayed on the CRT and printed.

4. MEASUREMENT STANDARDS

Calibration in the usual sense is not necessary. Signal response can be converted into a size distribution without standards. However, it is assumed that the instrument is aligned properly, the theory is properly applied, sampling errors are negligible, and detector response is spatially uniform, or nonuniform response is corrected in software. To test these assumptions, it is necessary to use standards.

The most generally available standards are latex particles.* Those below 2 μm are usually sized with an electron microscope. The uniformity from batch to batch is on the order of ± 1%, and the accuracy is probably no better than ± 5%, unless the specific batch received has been measured. Larger particles are typically sized by some form of microscopy. Batch uniformity is on the order of ± 5%, and accuracy is ± 10%. In addition, these larger sizes are not monodisperse. Relative standard deviations are on the order of 20%.

* Dow Diagnostic, Inc., P.O. Box 68511, Indianapolis, IN 46268; Duke Scientific Corp., 445 Sherman Ave., Palo Alto, CA 94306; Polysciences Inc., Paul Valley Industrial Park, Warrington, PA 18976.

Because latices aggregate, it is necessary to add surfactant and ultrasonicate before measurements are made (see Chapter 1, Figure 14). These steps are often neglected, and shifts toward somewhat larger average sizes are common. The larger standard sizes have broader distributions, and it is necessary to transform the number distribution statistics supplied by the manufacturer to weight distribution statistics before valid comparisons can be made. The smaller standard sizes have narrower distributions, thus this transformation is usually not significant.

Several authors have measured latices with a Malvern FD instrument (14, 17, 18). Measurements usually agree with expected results within $\pm$ 10%, if the precautions mentioned above are taken. Figure 7 shows measurements on 2.02-, 10.0-, and 43.9-μm latex standards. The agreement is very good, with the exception of the 2.02-μm standard. As explained in Section 6.1, this result is expected, since this size in water is not entirely within the FD regime.

Azzopardi (19) and Negus and Azzopardi (16) compared the FD results on glass beads with results using photographic and sieve analysis. At the time of their investigations, they did not have appropriate software for properly comparing multimodal distributions. For the best-characterized sample (the one with the largest number of particles photographically), the results are, in terms of Rosin–Rammler parameters:

Sieve	Photography	Fraunhofer
78 μm	73 μm	76 μm
5.2 μm	6.2 μm	6.0 μm

The only drop size generator commercially available that produces known monodisperse sizes without calibration is the Berglund–Liu generator, which is available from Thermo Systems, Inc., Although the concentration of drops from this instrument is generally too low for the FD instrument to get an adequate signal, Kennedy (20) was able to measure several very narrow distributions from 15 to 35 μm by allowing the drops to concentrate along the laser beam. Figure 8 shows the excellent agreement. Kennedy used the Rosin–Rammler software to fit the data; however, a different treatment of the FD pattern for these very narrow distributions is more appropriate and yields answers much faster (21).

5.   APPLICATIONS

Since the initial paper by Swithenbank et al. (8), a variety of applications have been demonstrated. Those described in an earlier review (22) include

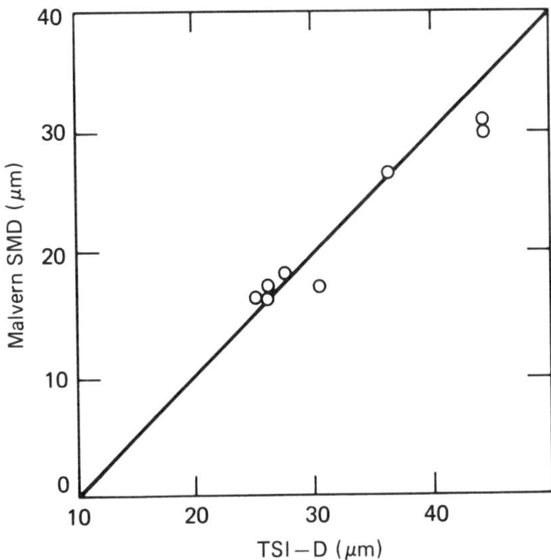

Figure 8. Comparison of Malvern Fraunhofer diffraction results and Thermo Systems, Inc. vibrating orifice aerosol generator: SMD is the Sauter or area average diameter; D is the diameter of the monodisperse aerosol from the TSI generator. Used with permission from reference 20.

measurement on radioactive, plastic, tracer particles used in blood flow measurements; ink particles used in copying machines; and characterization of aerosol can products. Felton (17) measured several samples of solids in water. These include latex spheres, zirconia fibers, Al_2O_3, glass ballotini, and Ni–Mg microballoons. Felton (23) also analyzed in-line holograms of developing drop size distributions from an electronic fuel injector, using the FD instrument to reconstruct and analyze the size distributions captured by the holographic technique. Since the resolution time of this technique is approximately 10 nsec, the effective resolution time of the FD instrument, used in this off-line mode as a detector, has been significantly reduced.

Simmons and Harding (24) measured the Sauter Mean Diameter (SMD) of both water and kerosene atomized in six different simplex fuel nozzles of small flow capacity. The SMD is the volume-to-surface area average diameter used in characterizing fuel sprays. Figure 9 shows the relative SMD as a function of Weber number (ratio of inertial to surface tension forces), which is proportional to the pressure drop across the nozzle Δp_F. The results show that data from different nozzles operating under various conditions can be correlated quite nicely. Simmons and Harding conclude

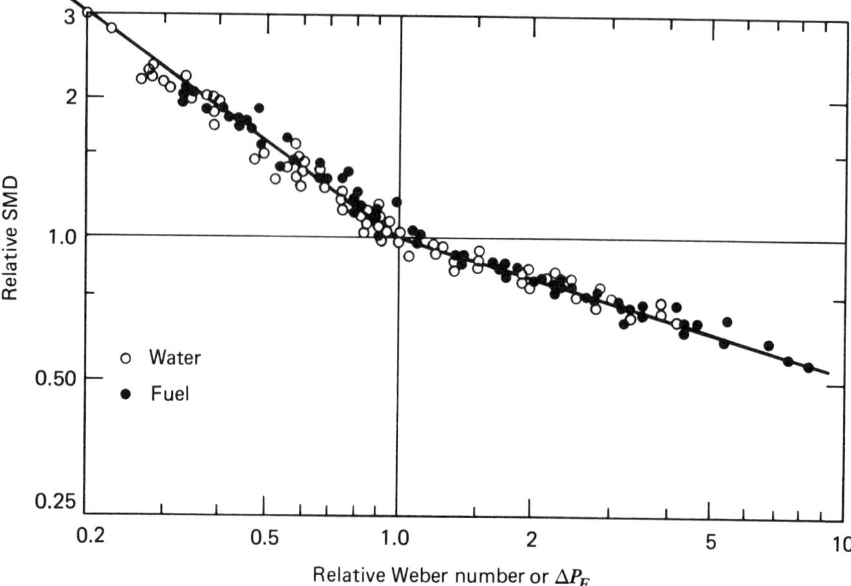

Figure 9. Correlation of Weber number with SMD; ΔP_F is the pressure drop across the fuel spray nozzle. Used with permission from reference 24 and the American Society of Chemical Engineers.

that the speed of the FD instrument, and consequently the large number of data points taken, was crucial in elucidating the effects of Weber number.

Another interesting application involving speed is the one by Felton (25) on the growth of KI crystals with time from a saturated solution that was seeded and allowed to cool slowly. Figure 10 shows $\bar{x}$ the average size parameter in the Rosin–Rammler (R–R) model, as a function of time and temperature. The exponent in the R–R equation (Equation 21) was constant throughout the measurement, indicating that the relative width of the size distribution was not changing.

6. LIMITATIONS

From the previous discussion and references, it is obvious that FD is a very useful and widely used technique for sizing particles. However, the complete scope of particle sizing problems is so vast that no single instrument is applicable in all situations. It is, therefore, helpful to enumerate the limitations of this technique. Also, many of the these limita-

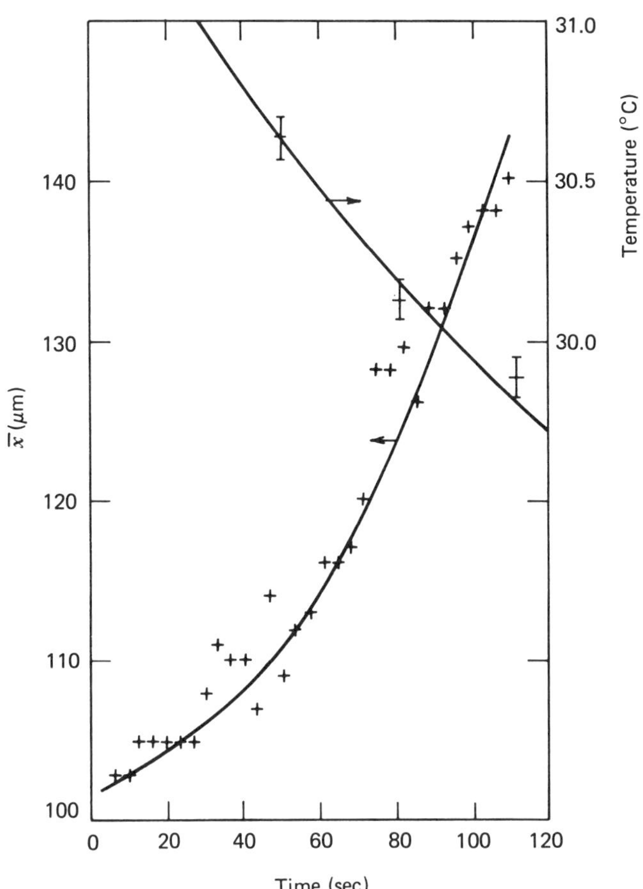

Figure 10. Growth of KI crystals from a saturated solution. Used with permission from reference 25.

tions apply equally well to other particle sizing techniques. The user must often compromise between what measurements are needed and what can be accomplished.

6.1. Theoretical

As discussed in Section 2.4, applying FD theory to particles above 10 μm rarely leads to large errors. Between 1 and 10 μm, it is wise to consider the size of the error that may occur. This can be done by calculating the difference between Mie scattering and FD results (11). However, the calculation is not recommended for general purpose use, since it is difficult.

Instead, Equations 9 and 10 can be employed. For example, consider the three different latex samples described in Section 3. The number average diameters, as given by the manufacturer, are 2.02, 10.0, and 43.9 μm with relative standard deviations in the number distribution of 0.7, 6.6, and 5.7% respectively. Assuming that these narrow distributions follow the log-normal distribution, the average diameter of the volume distributions is the same or slightly higher: namely, 2.02, 10.13, and 44.3 μm. Using the midpoints of the data shown in Figure 7, the average diameters of the volume distributions are 2.64, 11.46, and 48.49 μm. The differences between the FD results and the corrected values from the manufacturer are 30, 12, and 9.5%, respectively.

The FD sizes are larger, and some of the differences may be caused by aggregation. Also, better resolution is obtained for the two larger latices if larger focal length lenses are used. Keeping in mind that the accuracy of the manufacturer is probably only $\pm$ 5–10%, it is clear that something is wrong with the FD result for the 2-μm sample.

The relative index of refraction of these samples is $m = 1.60/1.33 = 1.20$. With $\lambda = 0.6328$ μm, Equation 9 gives $p_2 = 4.0$, $p_{10} = 20$, and $p_{43.9} = 87$. According to Equation 10, the 2-μm sample in water is not in the FD regime; whereas, the 10-μm sample is close to the FD regime and the largest size is within the FD regime. These predictions are confirmed by the measurement.

Sometimes, it is possible to choose the media in which the particle is suspended. For instance, if the 2-μm latex is suspended in air instead of water, then $p_2 = 12$ instead of 4 and the error is less.

This theoretical limitation is not restricted to the instrument described in Section 3, but is common to all FD instruments. Also, the errors described here are absolute or systematic errors. Precision, speed of measurement, and convenience are not affected. New software that provides, to a first order, a correction for the anomalous diffraction described in this section should be available soon.

6.2. Concentration

In analyzing the data, it is tacitly assumed that parallel, incident light is diffracted only once before it is registered on the detector. Multiple scattering, owing to high concentration, is not accounted for in the equations. In general, multiple scattering theory is not well-developed in any field of light scattering. Therefore, it is important to arrange for "dilute" systems. How dilute is dilute? Fortunately, there are some guidelines from experiment and theory.

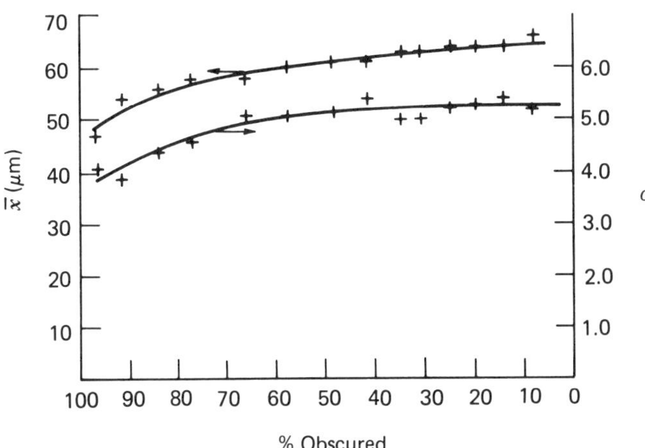

Figure 11. Variation of Rosin–Rammler parameters with obscuration for 45.5-μm Dow latex spheres (Equation 21). Used with permission from reference 17.

Figure 11 shows the results for one of five different samples run by Felton (17) as a function of concentration. The Rosin–Rammler size parameter is plotted on the left ordinate, and the R–R exponent σ is plotted on the right ordinate. The abscissa is percent obscuration ($100 - \%T$). The crosses represent pairs of Rosin–Rammler parameters at different concentrations with their corresponding transmittances.

It is clear that there is a slight shift to lower average sizes ($\bar{x}$ decreases) and somewhat broader distributions (σ decreases) as concentration increases and transmittance decreases.* This trend is predictable. Multiple scattering results in larger diffraction angles, since light diffracted once is no longer parallel when it strikes another particle. With more diffracted light reaching the outer parts of the detector, it is as if the distribution contained smaller size particles. This explains the shift to smaller average sizes. However, the larger particles are still diffracting light at very small forward angles. Even with some multiple scattering, the light indicating the presence of these larger particles registers. Hence, the distribution appears broader, because it contains the large as well as the apparently smaller "particles."

As seen in Figure 11 the effect of concentration is slight, and up to about 50% transmittance, the systematic error is negligible. Thus, ex-

---

* For the Rosin–Rammler equation (Equation 21), as σ decreases, the line width of the distribution increases.

perimentally, the maximum concentration can be arbitrarily set to correspond to 50% transmittance.

If the concentration is too low, large random errors (low precision) occur. This is a result of having too few particles, and statistical variations predominate. Higher concentrations result in low random error (high precision); however, the systematic error (absolute error) increases. A convenient compromise is between 70 and 80% transmittance. At very high concentrations, the signal plus background may be lower than the background, because of light scattering away from the forward direction. This usually results in large biases toward small sizes and nonphysical results. This situation is easily anticiated experimentally when transmittances below 5%, typically, are measured.

It is possible to calculate estimates of concentration that are either too high or too low. This is useful in the design of experiments. Starting with Equations 4 and 5 for the turbidity and transmittance and assuming that beam extinction is due entirely to scattering of Fraunhofer particles (Q_{sca} = 2), the following relationship can be derived,

$$\pi \bar{r}^2 \, N_M \, l_p = 0.5 \tag{22}$$

where $\bar{r}$ is an average radius, N_M is the maximum number of particles per unit volume, and l_p is the path length along the laser beam over which particles are present during the measurement.

The path length may be the distance between parallel windows in a sample cell, or it may be the estimated length of the laser beam that intersects the spray or airborne particles. Equation 22 corresponds to a transmittance of $1/e$.

Table 2 shows the results for the maximum number of particles/cm^3 for various path lengths and various average particle diameters. These values compare quite favorably with other light scattering and imaging techniques (26). With FD, one can usually measure systems that are more concentrated by a factor of at least 10, as compared to other optical techniques.

The minimum concentration, N_m, may be roughly estimated from experience to correspond to about 99% transmittance. This changes Equation 22 into

$$\pi \bar{r}^2 \, N_m \, l_p = 0.005 \tag{23}$$

Within this approximation, Table 2 can also be used simply by dividing the entries by 100.

The concentration limits calculated from Equations 22 and 23 are only estimates. Since the concentration varies inversely as the square of the

Table 2.   Maximum Concentration, N_M (number/cm^3), for Three Common Path Lengths, l_p (mm), Versus Average Diameter, $\bar{d}$ (μm)

$\bar{d}$	l_p		
	10	100	300
1	6.4×10^7	6.4×10^6	2.1×10^6
2	1.6×10^7	1.6×10^6	5.3×10^5
5	2.6×10^6	2.6×10^5	85,000
10	6.4×10^5	64,000	21,000
20	1.6×10^5	16,000	5,300
50	26,000	2,600	850
100	6,400	640	210

radius, any error in estimating the average radius results in a large error in N_M and N_m. However, although concentrations 10 times higher or lower than the calculated limits result in large errors, concentrations within a factor of 2 of the desired limits may be acceptable. Only an actual experiment can provide the answer. Equations 22 and 23 are also useful for experimental design. If estimated concentrations are much too high, the sample should be diluted or the path length shortened.

6.3.   Vignetting

Depth of field for imaging is usually defined as the distance on either side of the object over which acceptable definition in the image plane is attained when the lens is focused on the object. This can be quite crucial for some techniques where depth of field is quite small. The edges of large particles may not be well-defined. However, in the FD technique, the detector is not in the image plane, but in the focal plane. So depth of field, in the usual sense, is not a problem.

However, if the particles are too far away from the receiving lens, it is possible that the diffracted light from the main beam will be cut off by the receiving lens' finite aperture. The smallest particles in a sample have the largest diffraction angle and are thus subjected to vignetting first, as the sample is moved farther and farther from the lens. It is easy to demonstrate this experimentally, and the result is biased toward larger sizes.

It is possible to calculate the farthest extent of the sample from the lens at which vignetting is just noticeable for the smallest size that the particular lens can "see" (27). This smallest size corresponds to the outer radius of the largest ring on the detector. Figure 12 illustrates the optical

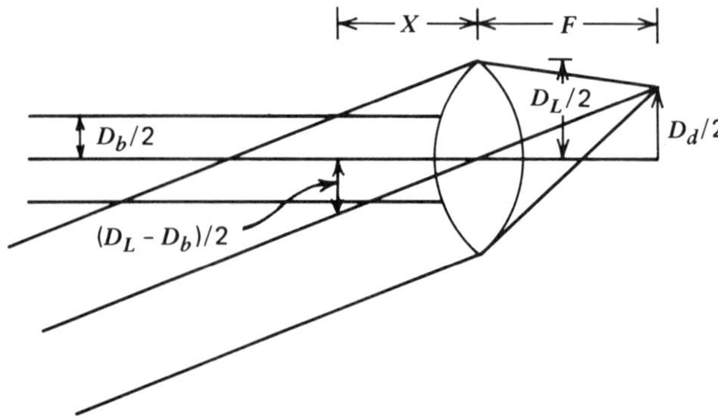

Figure 12. Optical dimensions for calculating vignetting: D_b is the expanded laser beam diameter; D_d is the diameter of the largest detector ring; D_L is the usable lens diameter; F is the focal length of the lens; x is farthest distance from sample to lens at which vignetting is noticeable.

arrangement for the calculation. The definitions and values in the figure are as follows:

D_b Laser beam diameter, 9 mm.
D_L Usable lens diameter; 35, 45, and 48 mm.
F Focal length of lens; 63, 100, and 300 mm, respectively.
D_d Outer diameter of largest ring, 28.6 mm.

Using similar triangles one finds,

$$x = \frac{F(D_L/2 - 4.5)}{14.3} \tag{24}$$

from which the following values are obtained:

F(mm)	x(mm)
300	409
100	126
63	57

For samples contained within cells, it is easy to push the cell close to the lens and satisfy the distance requirements given above. For airborne samples, care must be taken not to move the sample too far from the lens.

The criterion developed here applies to the smallest size for a given lens. It should be obvious that if that size doesn't exist in the sample to begin with, no bias will result. It is usually fairly easy to make measurements as a function of distance from the lens. When the average size begins to systematically increase, the maximum distance from the lens for that sample has been reached.

6.4. Data Transformation

Consider a single particle counter (SPC) where the detector response is proportional to the first power of the diameter. Assuming that sampling errors, including statistical errors, are negligible, the end result of a set of measurements is a set of N_i's and d_i's from which, in theory, all statistics and distributions can be calculated. However, a relative error of 10% in d_i becomes a 40% relative error in d_i^4. Thus, although in theory, data from an SPC can be transformed to calculate area and volume or weight statistics, the errors can quickly propagate.

The reverse transformation can also introduce large errors. If the detector response is proportional to the volume of the particle, then division by d_i^3 to get N_i can also lead to large errors. Therefore, if a choice exists, it is better to use an instrument whose response is directly proportional to the power of the diameter required.

Very few instruments have detector responses that are strictly proportional to an integer power of the diameter. Equation 17 shows that the FD instrument has a response that is proportional to the cross-sectional area, d^2, times a complicated function of d. However, the response is closer to the square or cube of the size than to the first power. For this reason, the data are most prudently analyzed in terms of area or volume or weight of particles in various class sizes, rather than in terms of number.

If the distribution by volume or weight is narrow enough, and has been well-resolved, then transforming to an area and number distribution is feasible, provided that the low end of the distribution has not been truncated. The opposite is true for number distributions. Transformations to higher powers may be acceptable, if the high end of the size distribution has not been truncated.

As an example, consider the following distribution: 10^3 particles of 1 μm, 10^2 particles of 10 μm, and 10 particles of 10^2 μm. Number, area, and volume average diameters are:

$$\bar{x}_N = 2.7 \qquad \bar{x}_A = 91 \qquad \bar{x}_V = 99$$

Suppose an instrument cannot measure the largest size range. If the 100-μm particles are truncated, the new averages become:

$$\bar{x}_N = 1.8 \qquad \bar{x}_A = 9.2 \qquad \bar{x}_V = 9.9$$

The number average has not changed much, the higher moments have.

The opposite case, where the lowest sizes are not included, gives the following biased results:

$$\bar{x}_N = 18 \qquad \bar{x}_A = 92 \qquad \bar{x}_V = 99$$

Here, the lowest moment is significantly changed.

The examples given above show that inadvertent extrapolation by data transformation beyond the domain of actual measurement can lead to erroneous results. Most instruments have restricted ranges in which they are useful. Data transformations can easily shift the results beyond the region where measurements were actually made.

7. AVERAGING

7.1. Ensemble

Some particle sizers count single particles. It is convenient to call such devices single particle counters (SPC's). When the technique used is optical, but not imaging, the device is conveniently referred to as a single particle optical counter (SPOC). Imaging techniques—photography, video, holography—are also obviously amenable to counting single particles; however, imaging is sufficiently different from these other techniques that it is in a class by itself. The third class of optical sizers are ensemble averagers (EA's), such as the FD instrument. Here, the sum of signals from all the particles measured during the experiment must be analyzed to yield size distribution information.

SPOC's are capable of giving "absolute" concentrations in terms of the number of particles per unit volume, if the volume element is well-defined. EA's are generally not capable of giving "absolute" concentrations. In particular, the FD instrument described here does not yield particle concentrations on an absolute basis.

EA's usually have an advantage over SPOC's in terms of the number of particles detected during a given time. Since they average over all the particles present, rather than count each, they are, in principle, faster than SPOC's, and they are less subject to statistical variations arising from registering too few particles.

7.2. Line-of-Sight

When spatial variations in particle size distribution are important, it must be recognized that the instrument is also averaging over a cylindrical region of space defined by the expanded laser beam diameter and path length of the beam in the sample. Spatial variations are usually not important in liquid-borne samples; however, they may be important in air-borne samples.

An instrument that measures in a region of space smaller than any spatial variations in size distribution gives a point measurement. Photographic techniques with limited depths of field are examples. The small region over which two small laser beams cross is another example.

Yule et al. (28) and Hammond (29) described how line-of-sight measurements can be deconvoluted to give point measurements. The deconvolution is called an Abel transformation and is most easily performed when airborne particles form an axisymmetric pattern. Yule's comparison with point measurements done with flash photography confirms the feasibility of this type of transformation.

In theory, knowledge of the size distribution as a function of position is useful. However, in many applications, such point measurements are combined to give a global average to characterize the overall size distribution. Also, point measurements are inherently slower to acquire than a global average over a large portion of the sample.

7.3. Velocity

As mentioned in Section 2.5, the FD pattern for any single particle does not depend on velocity, since the diffraction pattern is stationary. However, there is a difference in sampling if particles of different sizes have significantly different velocities. For liquid-borne samples, this is generally not the case. However, for airborne droplets and particles, it may be important.

Particles with slower velocities are weighted more heavily in the size distribution measured by the FD instrument described in this chapter, since they remain in the beam for a longer time. This effect can be made quantitative by considering more carefully what is meant by N_i, the number of particles in the ith class size. For instance, the number average size can be written as

$$\bar{x}_{1,0} = \frac{\sum x_i N_i}{\sum N_i} \tag{25}$$

where the sum is over M class sizes, each represented by size x_i.

The subscript notation refers to the power to which the size is raised in the numerator and denominator, respectively. It is a notation first suggested by Mugele and Evans (30), and it avoids confusion in naming the various types of size distribution averages.

With this notation, the weight average diameter is

$$\bar{x}_{4,3} = \frac{\sum x_i W_i}{\sum W_i} = \frac{\sum x_i^4 N_i}{\sum x_i^3 N_i} \tag{26}$$

where W_i, the weight of particles in the ith class size, is replaced by $x_i^3 N_i$. This quantity is proportional to W_i through the density. It is assumed that the density is constant for all sizes.

In many instruments, including the one described here, which are averaging over a region of space, N_i is the concentration in the usual sense with the following meaning:

$$N_i = \frac{\text{Number of particles in } i\text{th size class}}{\text{Unit volume of space}} = \frac{\#i}{m^3} \tag{27}$$

Other instruments, including many of the SPOC's, measure the flux of particles,

$$\mathbf{N}_i = \frac{\text{Number of particles in } i\text{th size class}}{\text{Passing unit area in unit time}} \hat{\mathbf{j}} = \left(\frac{\#i}{m^2 \text{ sec}}\right) \hat{\mathbf{j}} \tag{28}$$

where $\mathbf{j}$ is a unit vector in the direction of motion of the particles moving with velocity $\mathbf{v}_i$.

An SPOC that is also performing a point measurement is really measuring flux. The unit area is often the area of the opening of the inlet port of the probe used for sampling.

The American Society for Testing and Materials Subcommittee E29.04 on liquid particle characterization proposed the use of the term "spatial averaging" when N_i is measured and the term "temporal averaging" when $\mathbf{N}_i$ is measured. Although there is some confusion, because these terms have other meanings, it is helpful to distinguish between these two methods of averaging by using the superscripts S and T. With these designations, the relationship between concentration and flux is given by

$$N_i^S (\#i/m^3) \times v_i (m/\text{sec}) = N_i^T (\#i/m^2 \text{sec}) \tag{29}$$

where the velocity and flux vectors are replaced by their scalar values for simplicity.

The "spatially" averaged, weight average diameter measured by the FD instrument can, therefore, be expressed in terms of "temporally" averaged quantities as follows:

$$\bar{x}_{4,3}^{S} = \frac{\sum x_i^4 N_i^S}{\sum x_i^3 N_i^S} = \frac{\sum (x_i^4 N_i^T / v_i)}{\sum (x_i^3 N_i^T / v_i)} \tag{30}$$

From the equation given above, it is clear that if all particles are moving with the same velocity, there is no difference between "spatial" and "temporal" measurements. If there is a size-dependent velocity distribution, the two types of averaging yield different results.

As an example, consider the case where the velocity distribution is linear with size:

$$v_i = k x_i \tag{31}$$

Then,

$$\bar{x}_{4,3}^{S} = \frac{\sum x_i^4 N_i^S}{\sum x_i^3 N_i^S} = \frac{\sum x_i^3 N_i^T}{\sum x_i^2 N_i^T} = \bar{x}_{3,2}^{T} \tag{32}$$

Since the higher moments of a distribution are numerically larger than the lower moments, it follows that

$$\bar{x}_{4,3}^{T} > \bar{x}_{3,2}^{T} = \bar{x}_{4,3}^{S} \tag{33}$$

Thus, a "temporally" averaged, weight average diameter is larger than a "spatially" averaged, weight average diameter. Slower moving particles are weighted more heavily in the size distribution statistics from measurements with an instrument that "spatially" averages.

It is generally more difficult, however, to sample without bias from an airborne sample using an instrument that "temporally" averages. Since a representative sample of particles must enter the inlet port and be counted, the presence of the probe must not disturb the flow. The sampling velocity entering the probe must equal the carrier velocity past the probe. This is called isokinetic sampling. The presence of the probe does change the carrier gas streamlines in the vicinity of the inlet. Changes in acceleration are not easily followed by all size particles uniformly. For example, sampling velocities higher than isokinetic result in bias toward small particles, which can more easily follow changes. Some of the problems associated with isokinetic sampling are discussed by Hesketh (31).

7.4. Equivalent Sphere Diameter

Not all particles are spheres, but most droplets are. This shape minimizes surface free energy, provided that no external forces are present. When shape is important, it is necessary to use an image analyzer, because all other techniques assume an equivalent sphere. This point is discussed in Section 1.

The nature of the "equivalent sphere" depends on the technique. Equivalent spheres obtained in sedimentation instruments fall at the same rate as the actual particle that is sedimenting. Equivalent spheres in cascade impactors are aerodynamically equivalent spheres with a density of 1 gm/cm^3, which cascade along the same streamlines and impact on the same plates as the actual particles. Equivalent spheres in sieving devices correspond to those with a diameter equal to the opening of the mesh through which the actual particle just fits. Equivalent spheres in light scattering devices correspond to spheres that give the same signal response as the actual particle.

Since the various particle sizing techniques respond differently to shape and orientation, it is not surprising that the equivalent sphere diameter (ESD) varies according to technique. This is one of the reasons why a distribution of nonspherical particles may give different results with different techniques.

The FD instrument gives an FD-equivalent spherical diameter. If the particles are randomly oriented, the ESD represents an average over all orientations. As Janzen (32) pointed out, it would be satisfying to relate the ESD of a particular technique to one of the geometrically defined ESD's. These include the volume-ESD, $D_v = (6V/\pi)^{1/3}$, the orientation-averaged, projected area-ESD, $D_{\bar{a}} = (4\bar{a}/\pi)^{1/2}$, the surface area-ESD, $D_{SA} = (S/\pi)^{1/2}$, and several others. In these definitions, it is assumed that a measurement is made of the particles' actual volume (V), orientation-averaged projected area $(\bar{a})$, or surface area (SA), respectively. This measurement is then equated to the equation for the volume, projected-area, or surface area, respectively, of a sphere. Janzen showed experimentally that the diameters of distributions of carbon black flocs, using Lorenz–Mie theory to interpret extinction data from light scattering measurements, were most similar to D_v.

It is often assumed that the diffraction pattern of an irregularly shaped particle is that of its projected area; this author can find no theoretical justification for such an assumption. However, approximate diffraction calculations by Hirleman (33) on particles made of equal-sized spheres shows, in the cases he studied, that the projected area assumption is a reasonable approximation.

Diffraction calculations of irregular shapes are exceedingly difficult, and it is doubtful whether approximations will find general use. It may be possible to get some information from FD on simple, nonspherical shapes (ellipsoids, long rods, cubes, etc.), if it is assumed, a priori, that the shape is known and that the orientation is either constant or completely random. However, since the ensemble average of the diffraction pattern of even irregularly shaped particles is uniquely related to the size distribution, it is still extremely useful to relate effective size distribution to the diffracted light energy distribution.

Finally, it should be pointed out that all of the other techniques for particle sizing, with the exception of imaging, have similar limitations. For example, terminal velocity equations, which relate size to sedimentation properties, exist only for the most simple shapes. Exact solutions to the general Lorenz–Mie problem are limited to spheres and oriented ellipsoids (34).

8. SUMMARY

As discussed in Chapters 1 and 2 no one instrument can cover the large range of particle and droplet sizing applications. This is true not only because the size range required is so large, but also because the range of physical properties encountered is so vast. Some instruments, where applicable, are simple and inexpensive: sieving and manual microscopy are examples. Such devices are usually labor intensive and/or subject to operator bias. In addition, they are usually restricted in their applications. Automatic or semiautomatic instrumentation is less labor intensive and usually more costly, initially, such as the FD instrument described in this chapter. Its advantages and disadvantages may be summarized as follows:

1. The instrument, as configured, is very versatile. It can be used for size distribution analysis of particles or droplets suspended or flowing in any reasonably clear liquid, conducting or nonconducting, or when particles or droplets are airborne. The size range covered by all the lenses is broad. The computer and printer allow for a large choice in data analysis, manipulation, and presentation.

2. The measurement, like many optical measurements, is nonintrusive. There is no probe to disturb the flow and to introduce sampling errors.

3. In the range where FD theory is applicable, no calibration is necessary, results are independent of the refractive index, and accuracy is ± 5%. Below about 10 μm, however, a systematic error may occur. This

does not affect precision or selectivity, both of which are prerequisites for process control.

4. Repeatability is ± 3%. Resolution is good, provided the correct lens is used to make measurements. In the largest size ranges, resolution is only fair.

5. The instrument is quite easy to set up and to operate. There are no orifices to clog, no images to analyze, and no extensive calibration procedures to follow. Typically, measurements take a few seconds, and results are calculated in a few minutes.

6. Although it is not possible to use the technique when the transmittance is low, the range of concentration over which results are obtainable is competitive with most other techniques.

7. The instrument is not an SPC. It does not measure at a single point and, at the present time, it is not useful for absolute concentration measurements. It does, however, ensemble average over a large number of particles rapidly and over a region of space. In the majority of cases, this type of averaging is sufficient and often necessary.

8. Particle shape is not measured. Results are in terms of an orientation-averaged ESD that is probably most similar to the orientation-averaged, projected area ESD.

9. Measurements are biased toward slower moving particles, which is true for any instrument that is sensitive to the number of particles/unit volume of space rather than to flux. However, for particles moving at the same or nearly the same speed, which is true for the vast majority of liquid-borne particles, the effect is rarely signficiant.

10. The equivalent depth of field is very large. There is a vignetting problem which is only significant for the smaller sizes when using long path lengths far from the receiving lens.

11. In the data analysis, it is assumed that the density is the same for all particles, independent of size. For a mixture where this is not true, errors will arise. The same assumption is made in sedimentation techniques where Stokes' Law is invoked.

NOTATION

$\bar{a}$ Orientation-averaged, projected area.

d Particle diameter.

$\bar{d}$ Average particle diameter.

F Focal length of lens.

C	Detector optical constant.
C_{abs}	Total cross section for absorption, L^2.
C_{ext}	Total cross section for extinction, L^2.
C_{sca}	Total cross section for scattering, L^2.
$D_{\bar{a}}$	Orientation-averaged, projected area equivalent sphere diameter.
D_{sa}	Surface area equivalent sphere diameter.
D_v	Volume equivalent sphere diameter.
I	Intensity of the diffraction pattern.
I_0	Intensity of incident beam.
I_T	Intensity of transmitted beam.
$\hat{\jmath}$	Unit vector in the direction of motion of the particle.
J_0	Zero-order spherical Bessel function.
J_1	First-order spherical Bessel function.
k	Constant.
K	Constant.
l	Path length of cell.
l_p	Path length over which particles are present during measurement.
L	Fraction of light energy within a given circle on the detector.
L_{s_1,s_2}	Fraction of light energy falling between two radii, s_1 and s_2.
m	Refractive index of particle relative to medium in which it is suspended.
N	Number of particles per unit volume.
N_m	Minimum concentration of particles, number of particles per unit volume.
N_M	Maximum concentration of particles, number of particles per unit volume.
P	Van de Hulst's particle size parameter, $2\pi d \mid m - 1 \mid /\lambda$.
Q	Cross-sectional efficiency, $C/\pi r^2$.
r	Particle radius.
$\bar{r}$	Average particle radius.
s	Radial distance in the detection plane measured from the optical axis.
S	Superscript indicating spatial average.
SA	Average particle surface area.
T	Transmittance.
T	Superscript indicating temporal average.
v_i	Speed of a particle.

$\mathbf{v}_i$ Velocity vector of particle.

V Average particle volume.

W Weight of a particle.

x Size parameter of a sphere, $\pi d/\lambda$; argument of Bessel function, $2\pi rs/\lambda F$; independent variable in model distribution functions.

$\bar{x}$ Size parameter in model distribution function.

$\bar{X}_A$ Area average particle diameter.

$\bar{X}_N$ Number average particle diameter.

$\bar{X}_V$ Volume average particle diameter.

λ Wavelength of light in the suspending medium.

θ Angle normal to incident beam.

ρ Particle density.

τ Turbidity, L^{-1}.

σ Width parameter in model distribution functions.

REFERENCES

1. T. Allen, *Particle Size Measurement*, 3rd ed., Chapman & Hall, New York, 1981.

2. B. H. Kaye, *Direct Characterization of Fineparticles*, Wiley, New York, 1981.

3. J. D. Stockham and E. G. Fochtman (Eds.), *Particle Size Analysis*, Ann Arbor Science Publishers, Michigan, 1978.

4. I. Chabay and D. Bright, *Chemtech*, **9**(11), 694–699 (1979).

5. J. D. Stockham and E. G. Fochtman (Eds.), *Particle Size Analysis*, Ann Arbor Science, Ann Arbor, 1978, p. 30.

6. B. H. Kaye, *Direct Characterization of Fineparticles*, Wiley, New York, 1981, Chapter 10.

7. J. Cornillaut, *Appl. Opt.*, **11**, 215–218 (1972).

8. J. Swithenbank, J. M. Beer, D. S. Taylor, D. Abbot, and G. C. McCreath, *Prog. Astronaut. Aeronaut.*, **53**, 421 (1977).

9. M. Kerker, *The Scattering of Light and Other Electromagnetic Radiation*, Academic Press, New York, 1969.

10. H. C. Van de Hulst, *Light Scattering by Small Particles*, Wiley, New York, 1957.

11. A. R. Jones, *J. Phys. D: Appl. Phys.*, **10**, 138–140 (1977).

12. M. V. Klein, *Optics*, Wiley, New York, 1970.

13. R. A. Dobbins, L. Crocco, and I. Glassman, *Am. Inst. Aeronaut. Astronaut.*, **1**, 1882–1886 (1963).

14. Dr. David Siegel, Procter & Gamble Co., Miami Valley Laboratories, Ohio, private communication.

15. J. M. Tishkoff and C. K. Law, *Eng. Power,* **99,** 684–688 (1977).

16. C. Negus and B. J. Azzopardi, "The Malvern Particle Size Distribution Analyser: Its Accuracy and Limitations," United Kingdom Atomic Energy Establishment Report 9075, AERE, Harwell, England, 1978.

17. P. G. Felton, "In-Stream Measurement of Particle Size Distribution," paper presented at the International Symposium on In-Stream Measurements of Particle Solid Properties, Bergen, Norway, August 22–23, 1978.

18. D. C. Hammond, "Accuracy Verification of a Malvern ST1800 Analyzer," General Motors Research Publication GMR-3195, GM Research Laboratories, Warren, Michigan, 1980.

19. B. J. Azzopardi, "The Analysis of Malvern ST1800 Output by Different Models and a Comparative Test of a Near Monodisperse Distribution," United Kingdom Atomic Energy Establishment Report M3067, AERE, Harwell, England, 1980.

20. Dr. J. Kennedy, United Technology Research Company, East Hartford, Connecticut, private communication.

21. J. Swithenbank and D. S. Taylor, "Size Distribution Measurement for Near Monodisperse Particles and Sprays," Report No. HIC 315, Department of Chemical Engineering and Fuel Technology, University of Sheffield, England, February 1979.

22. B. Weiner, *Soc. Photoopt. Instrum. Eng.,* **170,** 53–62 (1979).

23. P. G. Felton, "Measurement of Particle/Droplet Size Distributions by a Laser Diffraction Technique," Department of Chemical Engineering and Fuel Technology Report, University of Sheffield, Sheffield, England, 1981.

24. H. C. Simmons and C. F. Harding, "Some Effects of Using Water as a Test Fluid in Fuel Nozzle Spray Analysis," Paper 80-GT-90 presented at the Gas Turbine Conference, New Orleans, Louisiana, March 10–13, 1980.

25. P. G. Felton and D. J. Brown, "Measurement of Crystal Growth Rates by Laser Diffraction," Paper presented at the Sixth Annual Institute of Chemical Engineers Research Meeting, University College London, April 4–6, 1979.

26. W. A. Cassatt and R. S. Maddock (Eds.), "Aerosol Measurements," National Bureau of Standards Publication 412, U.S. Government Printing Office, Washington, D.C., p. ix, 1974.

27. Dr. Lee Dodge, Southwest Research Institute, San Antonio, Texas, private communication.

28. A. J. Yule, C. Ah Seng, P. Felton, A. Ungut, and N. A. Chigier, "A Laser Tomographic Investigation of Liquid Fuel Sprays," 18th Symposium (International) on Combustion, The Combustion Institute, pp. 1501–1510, 1981.

29. D. C. Hammond, "Deconvolution Technique for Line-of-Sight Optical Scattering Measurements in Axisymmetric Sprays," General Motors Research

Report GMR-3361-R, General Motors Research Laboratories, Warren, Michigan, 1981.

30. R. A. Mugele and H. D. Evans, *Ind. Eng. Chem.*, **43**, 1317–1324 (1951).

31. H. E. Hesketh, *Fine Particles in Gaseous Media*, Ann Arbor Science Publishers, Michigan, 1979, pp. 151–152.

32. J. Janzen, *Appl. Opt.*, **19**, 2977–2984 (1980).

33. E. D. Hirleman, *Opt. Eng.*, **19**, 854–860 (1980).

34. D. W. Schuerman (Ed.), *Light Scattering by Irregularly Shaped Particles*, Plenum Press, New York, 1980.

PARTICLE SIZE MEASUREMENTS FROM 0.1 TO 1000 μm, BASED ON LIGHT SCATTERING AND DIFFRACTION

PHILIP E. PLANTZ

Leeds and Northrup Instruments
MICROTRAC Products
St. Petersburg, Florida

1. INTRODUCTION

Knowledge of particle size and particle size distribution in the range of 0.1–1000 μm is fundamental to a wide variety of industrial processes, including grinding, crystallization, emulsification, and polymerization. Such information is valuable in the production of particles of specific sizes to control process efficiency and product quality. In addition, advances in process technology have demonstrated the necessity for rapid and automated particle size measurement to keep pace with energy and labor costs and profitability demands.

The implementation of Mie theory with Fraunhofer diffraction and 90° (right-angle or side) scatter permits the measurement of particle sizes over

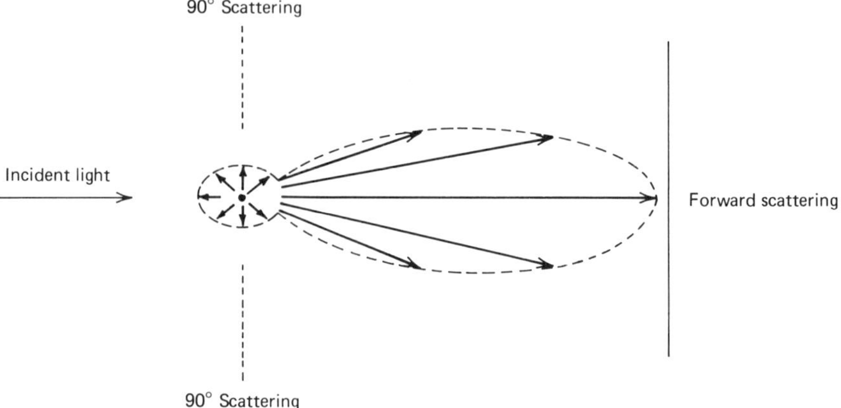

Figure 1. Angular distribution of light scattered from a single particle. The length of the arrow corresponds to the intensity of light in that direction.

the range of 0.1–1000 μm (1–7). Although light scattering was used previously for particle size measurements (8), relatively recent technological advances have permitted the development of practical, precise instrumentation that provides timely operator-independent measurements of wet or dry materials which, as a stream or cloud of flowing particles, are illuminated by a light source.

This chapter describes the theories leading to the development of MICROTRAC® Particle Size Analyzers, manufactured by Leeds and Northrup Instruments, St. Petersburg, Florida, while emphasizing applications of the instrumentation and correlation with other methods. (MICROTRAC® is a registered trademark of Leeds and Northrup Co., North Wales, PA.)

2. THEORETICAL CONSIDERATIONS

2.1. MIE Scattering

As particle size increases from 1/10 to 10 times the wavelength of the incident light, scattering from different portions of a single particle is out of phase, causing interference and reduced intensity. As shown schematically in Figure 1, phase differences are small for small scattering angles, causing a greater intensity of scattered light at low angles.

The overall effect gives rise to an angular distribution of scattered light (depending on refractive index and particle size) in which most scattering

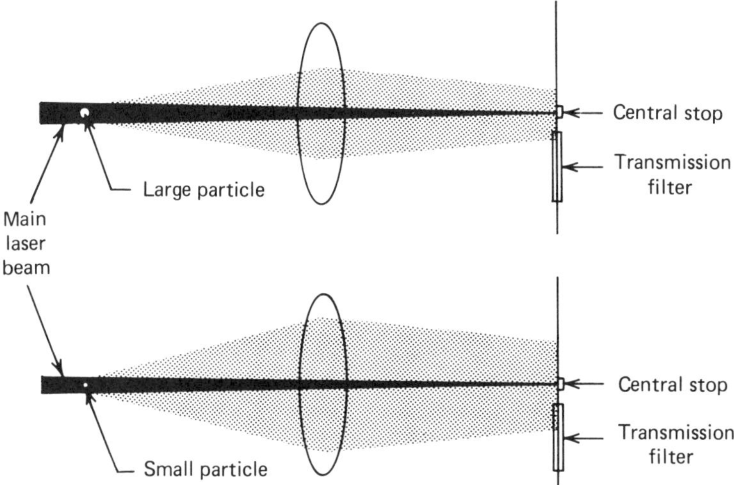

Figure 2. Comparison of scattered light flux angle and intensity for large and small particles. The central stop blocks light proportional to the fourth power of the particle diameter. The transmission filter (mask) permits scattering light proportional to the third power (volume) of the particle diameter to pass for measurement. Reprinted by permission of American Ceramic Society.

(greatest intensity) is in the forward direction. Thus, light collection in a plane intersecting the forward scattered light is sufficiently intense and contains appropriate information for the determination of particle size. For particle sizes approaching the wavelength of the incident light (0.63 μm for He–Ne laser), and to account for the effects of refractive index, rigorous Mie theory computations (2) must be employed.

2.2. Fraunhofer Diffraction

For applications of particles having a diameter much larger than the wavelength of light, there is little effect of refractive index and a special case of Mie scattering theory, Fraunhofer diffraction, is invoked (see Chapter 5). This theory explains that the intensity of light scattered by particles is proportional to particle size, and the size of the diffraction pattern (scattering angle) is inversely proportional to particle size. Figure 2 shows examples of light scattered from two different size particles.

Very close to the center of the pattern, the magnitude of the light intensity is proportional to the fourth power of the particle diameter, and because of such high intensity, is blocked. Smaller particles scatter a small, definite amount of light through a fixed, but larger, angle. Conversely, the larger particle scatters a greater amount of light, but through a smaller angle. For a single particle, the intensity distribution, $I(w)$, of the Fraunhofer diffraction pattern is given by the Airy equation:

$$I(w) = Ek^2x^4\left[\frac{J_1(kxw)}{kxw}\right]^2 \tag{1}$$

where E is the flux per unit area of the incident beam, $k = 2\pi/\lambda$, $w = \sin\theta$, and J_1 is the first-order Bessel function of the first kind. The dimensionless parameter, x, is defined as:

$$x = \frac{\pi d}{\lambda} \tag{2}$$

where d is the particle diameter and λ is the wavelength of the incident beam.

From this relationship, a third power or volume response of scattered light can be defined by a spatial filter having a well-defined mask (3). Figure 3 depicts one possible manner of spacing a series of such masks to observe a series of particle sizes separated into several channels. From this separation of the angular distribution of light, a histogram of the total particle size distribution is developed. The upper channel edges for one particular instrument range are 2.8, 3.9, 5.5, 7.8, 11, 16, 22, 31, 44, 62, 88, 125, and 176 μm. By a simple interchange of optical components, the channel edge may be changed to 4.7, 6.6, 9.4, 13, 19, 27, 38, 53, 75, 106, 150, 250, and 300 μm.

2.3. Side (90°) Scattering

For particles smaller than the light source wavelength, the forward scattered light flux (because of the wide angular distribution) is difficult to collect with a conventional optical system; thus, forward scatter is useful to approximately 0.5 μm, but particle sizes smaller than 0.5 μm are determined by measuring 90° scatter at three different wavelengths and two planes of polarization of each wavelength. Although the refractive index affects the relationship between forward and 90° scatter, compensation is made by system programming.

Measurements of particle size at 90° are practical, because of the effect of polarized light on scattering intensity. Scattering from particles illuminated by light polarized in one plane has a different intensity profile than that of light polarized in the orthogonal plane. To illustrate this point,

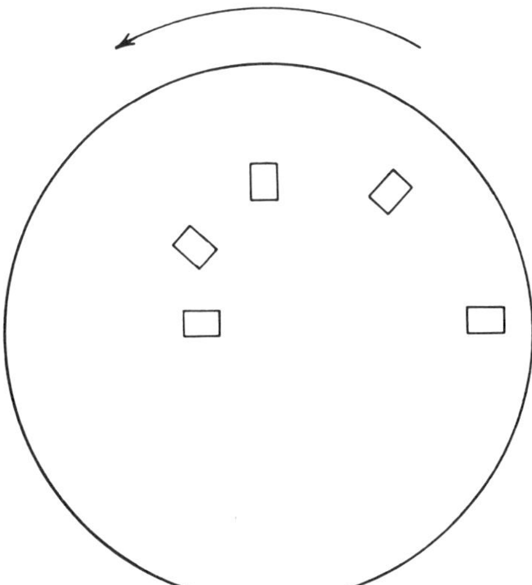

Figure 3. Five third-power masks spaced according to the angle of diffraction. An incident beam of light is focused at the center of the disk. Rotation of the mask permits sequential measurement of five particle size channels as diffraction pattern intensities, $I(w)$. MICRO-TRAC measures thirteen channels. Reprinted by permission of IEEE from *IEEE Trans. Ind. Appl.*, **15**(3), 323 (1979), copyright 1979 IEEE.

a plot of scattered light intensity from droplets suspended in air (relative refractive index, 1.33) as a function of angle for several x values can be developed as shown in Figure 4. For low x values (particle size and wavelength similar), the difference between the two scattered intensities, i_1 and i_2, is greatest at 90°. If the intensity difference $(i_1 - i_2)$ at 90° is divided by the total particulate volume and is plotted as a function of x, a peak occurs at 1.5 (Figure 5). Substituting in Equation (2),

$$1.5 = \frac{\pi d}{\lambda}$$

$$0.5 = \frac{d}{\lambda}$$

which shows that the maximum flux difference is obtained for water droplets having particle size one-half the wavelength.

This discussion has so far applied to single particles or monodisperse distributions of particles spherical in shape. Generally, real distributions of polydisperse, irregularly shaped industrial process particles are not

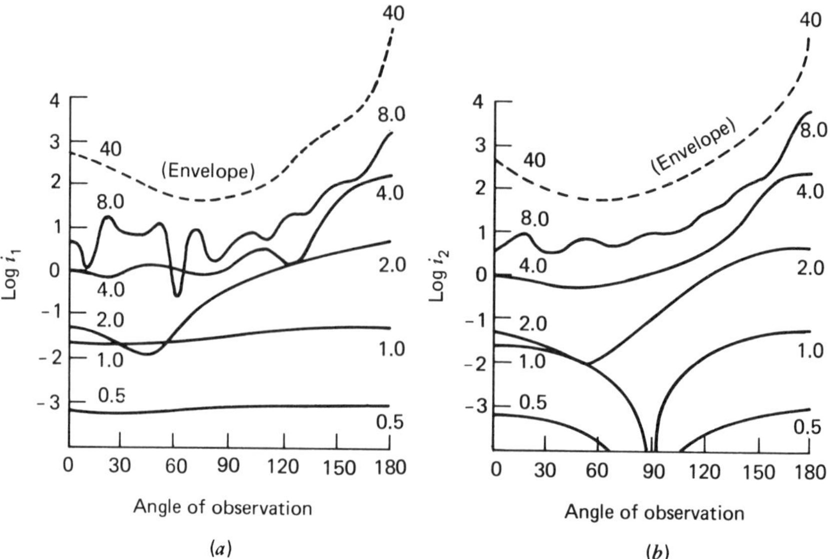

Figure 4.   Intensity of light as a function of angle for several x values for orthogonal planes of polarizations 1 and 2. (*a*) Polarization 1: intensity of i_1 for values of x from 0.5 to 40. (*b*) Polarization 2 (orthogonal): intensity of i_2 for values of x from 0.5 to 4.0. Reprinted by permission of Cahner Exposition Group.

greatly affected by physical properties such as index of refraction. For example, in particles having a relative refractive index of 1.6, the maximum flux difference per particulate volume corresponds to a wavelength (λ) of 0.9 μm (i.e., $d = 0.3$ μm).

Table 1 lists the maxima of the response curves ($i_1 - i_2$) for particulate materials generally available in many applications. From this data, we see that the amount of particles having a diameter of 0.15 μm can be determined at 90° to incident light ($\lambda = 0.45$ μm) at two orthogonal planes of polarization.

Table 1.   Wavelengths of Band-Pass Filters Useful for Monitoring Particles of Diameter d at 90° to the Incident Light

	Filters		
	Blue	Yellow	Infrared
$\lambda(\mu m)$	0.45	0.60	0.90
$d(\mu m)$	0.15	0.20	0.30

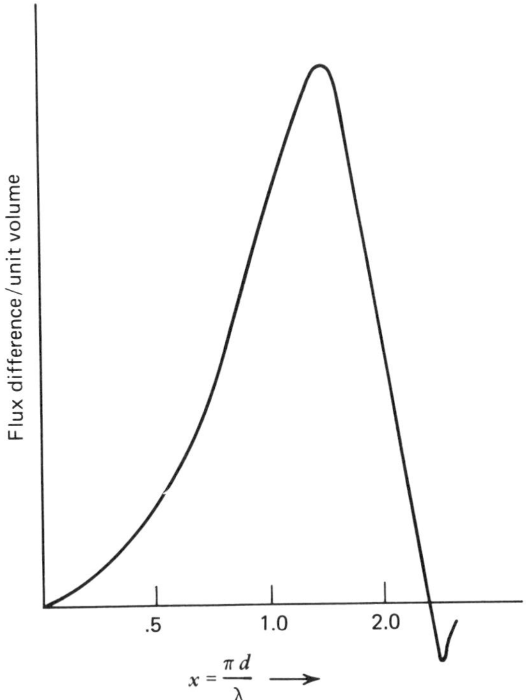

Figure 5.   Diagram showing maximum flux difference ($i_1 - i_2$) per unit volume as a function of dimensionless parameter x. Reprinted by permission of Cahner Exposition Group.

3. DATA PRESENTATION

Use of state-of-the-art microprocessors permits the collection of light flux data and the computation of particle size distributions, including specific surface, volume mean diameter, and 90th, 50th (median), and 10th percentiles for recording soft (computer) or hard (built-in printer) copies, as shown in Figure 6.

4. IMPLEMENTATION AND APPLICATION OF MICROTRAC ANALYZERS

4.1. Standard Range Analyzer

This analyzer measures particles over the range of 1.9–176 μm. Since the wavelength of the helium–neon laser is 0.63 μm and the particle size of

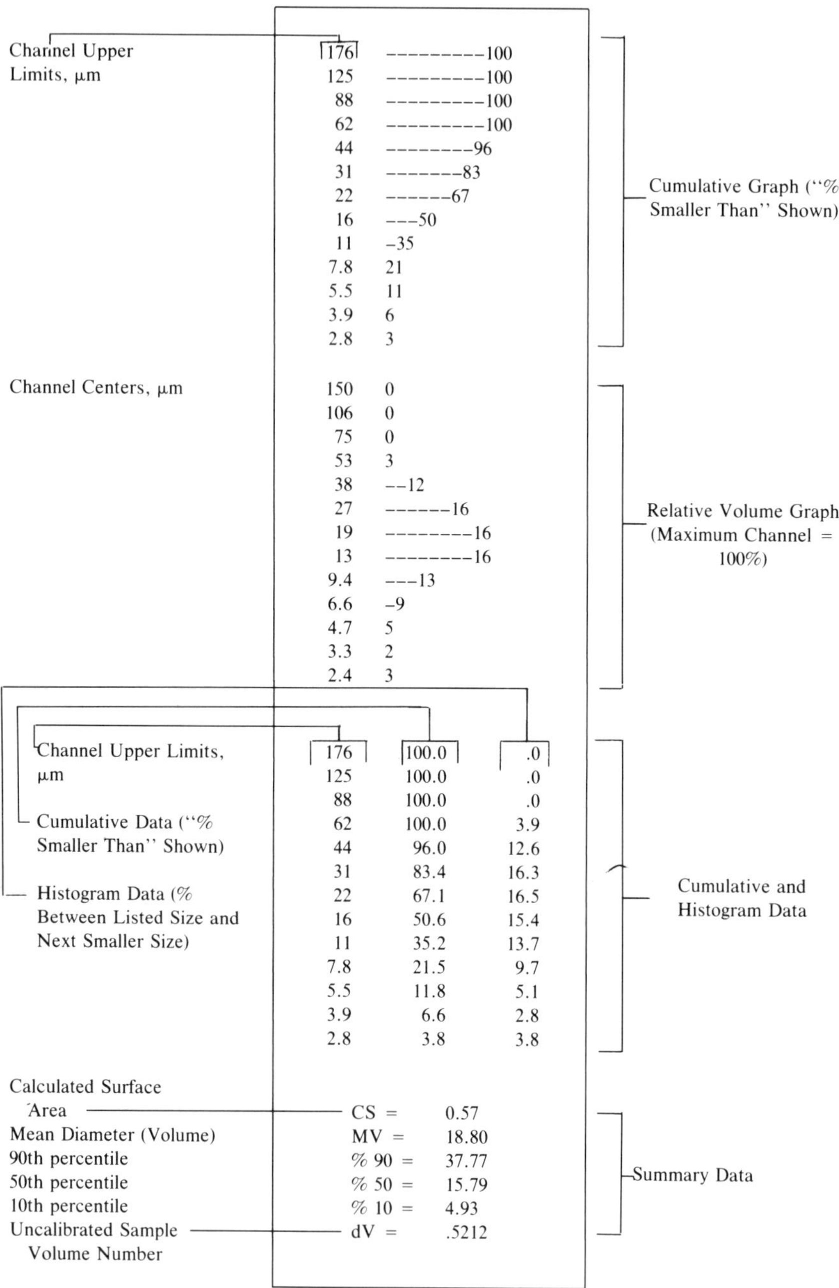

Figure 6. Typical data printout from MICROTRAC standard range (1.9–176 μm) particle size analyzer.

180

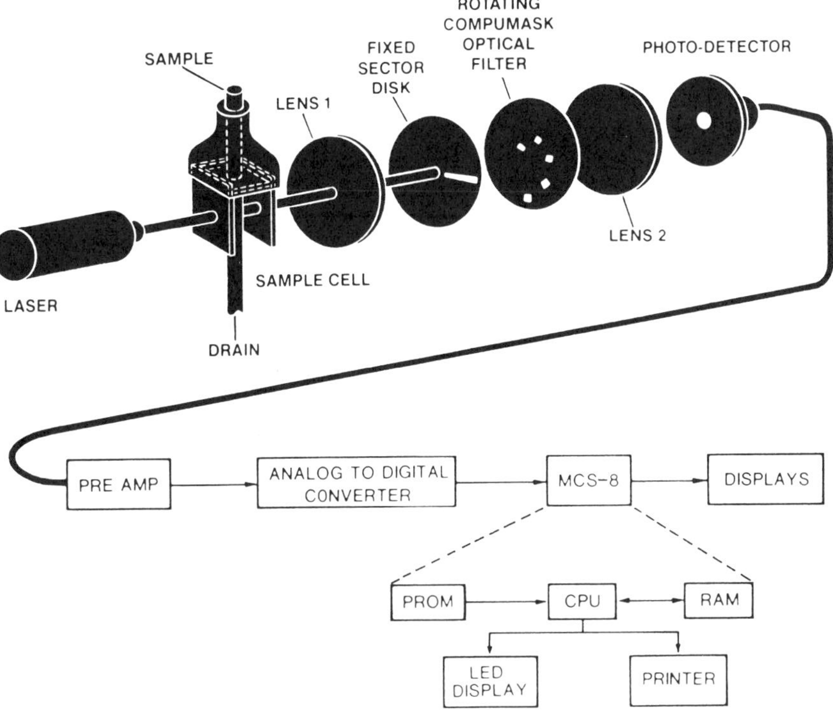

Figure 7.   Components of MICROTRAC standard range particle size analyzer. From reference 11. Reprinted by permission of the Society of Photo-Optical Instrumentation Engineers.

interest is much larger than the wavelength, Fraunhofer diffraction is employed as the scattering theory. As shown in Figure 7, light is scattered in the forward direction by a cloud of particles, giving rise to a "combined" diffraction pattern that is collected and focused by a lens onto the Compumask™. The Compumask™ contains a rotating disk that separates the pattern into 13 distinct zones or channel sizes by means of the 13 masks. Figure 3 shows 5 of the 13 masks. Light passes through each of the 13 masks sequentially, while the intensity is measured by a photodetector which transmits the signal to a microprocessor for storage. At the completion of each measurement cycle (nominally 100 sec), appropriate computations are completed for reporting the particle size distribution.

Fraunhofer diffraction assumes that index of refraction effects are inconsequential over an entire particle size range when using a source of a given wavelength. As mentioned, refractive index effects become more apparent as the particle diameter approaches the wavelength of illuminating light. The particle size at which the effects become apparent may be approximated (9) by

$$d = \frac{4.09\,\lambda}{2\pi\,|\,M - 1\,|}$$

where M = sample particle refractive index/refractive index suspending medium, λ is the incident wavelength, and d is the diameter.

The constant 4.09 is the maximum at the first-order effect of refractive index, as shown by the extinction curve for nonabsorbing spheres (9). Employing this equation, it can be shown that the refractive index of sample particles 1.9 μm in diameter should be 1.62 or greater for compliance with Fraunhofer theory; as particle size increases to 10 μm, the refractive index requirement decreases to 1.35, or very near the value of water. Again, this discussion pertains to monodisperse, nonabsorbing (transparent) spheres and can be considered to be a worst case.

In a more practical sense, the refractive index of most industrial materials is equal to or greater than 1.55 (M = 1.16), suggesting that refractive index effects begin at 2.4 μm, the very bottom portion of the measuring range of standard MICROTRAC. Also, since most industrial materials are irregularly shaped polydispersions, the index of refraction has little or no measurable effect.

As appropriate, samples may be measured in water, air, pure solvents, or media containing several components. For materials that neither dissolve in nor react with water, a 4-liter circulating system is included as an integral part of MICROTRAC instruments (Figure 8). The sample is premixed with water or water/surfactant and poured directly into the basin. Samples not requiring surfactants may be transferred directly to the basin, where suspension and dispersal are maintained by the circulation system, which transfers the suspended sample to the cell where light scattering occurs. The sample is continually recirculated during the measurement, until automatic drain–refill controls are actuated by the operator.

Flowing systems often contain a particulate velocity distribution where particles at the edge of the stream move more slowly than those at the center. The MICROTRAC circulating system circumvents this effect in part by a cell design that imparts turbulent flow, resulting in a well-mixed, homogeneous sample passing across the laser beam. In addition, the laser is directed to the center of the flowing medium, where velocity is maximum and the velocity distribution is of negligible effect. The combined

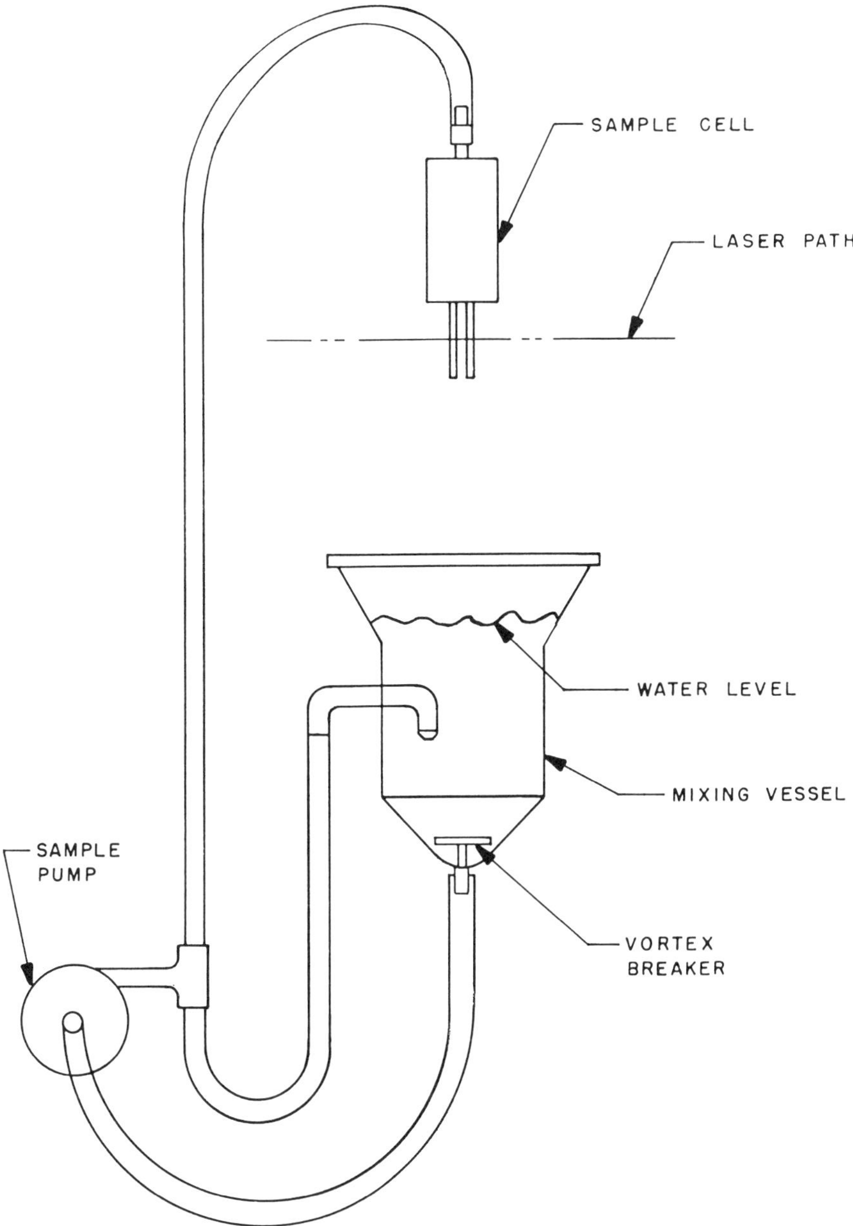

Figure 8. Schematic presentation of the 4-liter circulating system. The pump draws the sample from the mixing basin for transfer to the sample cell, from which the sample is returned to the basin. From reference 11. Reprinted by permission of Society of Photo-Optical Instrumentation Engineers.

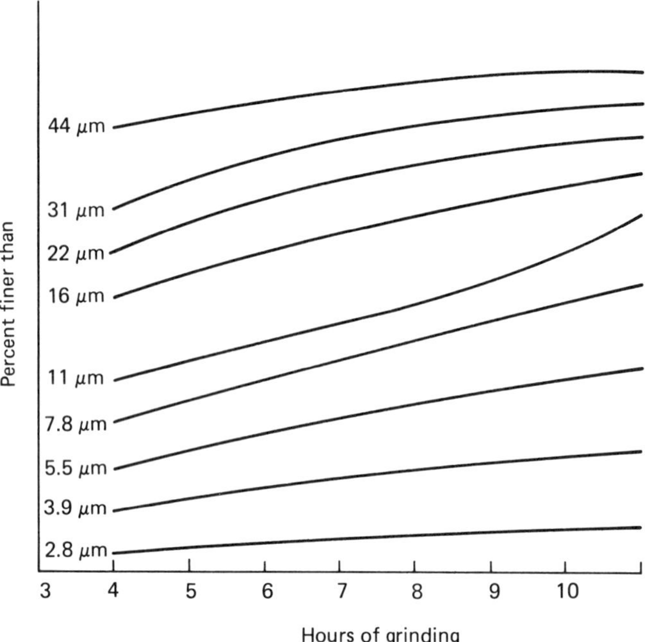

Figure 9. Change in particle size as a function of grinding time for flint media. Numbers to the left of the curves are particle sizes in microns. Positively sloped curves demonstrate increased fineness. Reprinted by permission of American Ceramic Society.

effects of cell design (turbulent flow) and laser positioning in the cell therefore avoid any flow bias.

This circulating system is useful in a wide variety of applications including aluminum, powdered alumina, coal powder, cosmetics, drilling muds, copper ore, taconite, soils, resins, pharmaceuticals, and ceramics (10–14). In many applications, the particle size distribution of raw materials is determined prior to use or grinding. Figure 9 shows particle size data plotted for a ceramic material ground for 3–10 hours. Increased fineness is determined by the constantly increasing slope of several size ranges. In this example, sample preparation required 30 sec and measurement 1 min, thus permitting real-time determination of particle size distribution in the grinding operation.

In field studies of soil and fluvial sediments (15), the light scattering method was compared to the glycerol method of surface area determination described by Kinter and Diamond (16). This method requires selective adsorption of a glycerol monolayer. Increase of particle weight, as a result of glycerol adsorption, is indicative of the number of glycerol

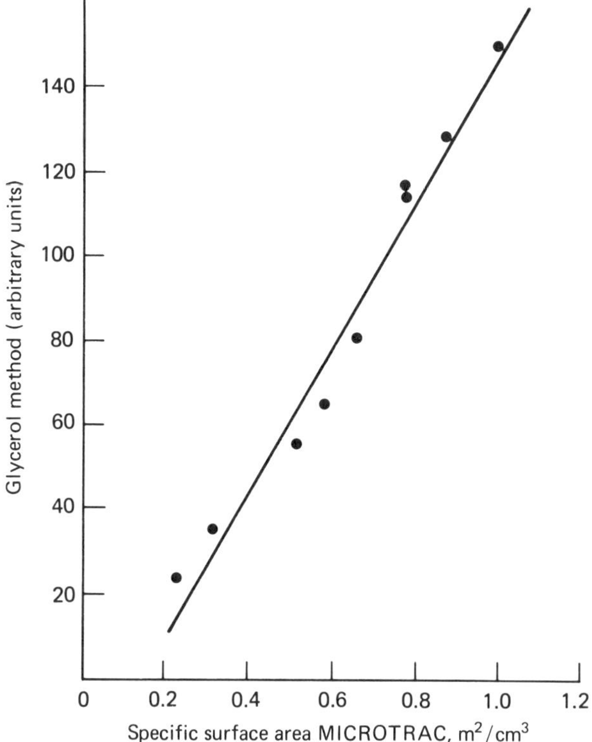

Figure 10.   Comparison of data obtained by MICROTRAC particle size analyzer and glycerol sedimentation method for several Arizona soil samples (from reference 15). Reprinted by permission of L. R. Cooper, U.S. Department of Agriculture, Agricultural Research Western Branch.

molecules adsorbed from which the surface area can be readily calculated. Computation of specific surface by MICROTRAC is performed from the volume distribution and is expressed as m^2/cm^3. As shown in Figure 10, the two methods show good correlation ($r^2 = 0.93$).

where X represents MICROTRAC data and Y represents pipette data.

In many applications, particulate materials not suspendable or wettable in an aqueous environment are easily dispersed in organic solvents. Such is the case with chocolate, where it is desirable to measure the combined particle size of solids excluding fat. For dispersal of chocolate, trichloroethane is employed in conjunction with ultrasonic energy. MICRO-TRAC accommodates the use of this solvent by means of a patented small volume recirculator (SVR) which is manufactured of anodized aluminum coated with Teflon for excellent chemical resistance. The SVR requires approximately 225 ml of sample or diluted sample, which is delivered to a glass cell by means of the pumping action of the SVR. During the measurement period, particles pass through the laser beam and diffract the light according to Fraunhofer theory, resulting in the particle size distribution.

To maintain the texture or "mouth-feel" portion of the chocolate flavor profile, manufacturers continually monitor the particle size distribution. Of special importance is the presence of particles larger than 10–15 μm. Particulates greater than this size cause a grainy "mouth-feel," resulting in low flavor scores. The SVR containing trichloroethane is used for measurement after preparing chocolate by suspension in trichloroethane and ultrasonic treatment. The solvent dissolves cocoa lipids, but does not affect the particle size of solvent-insoluble components such as sugars and milk solids. An example of this measurement technique is shown in Figure 11, where MICROTRAC laser light scattering data are compared to those from the Coulter Electronics conductometric measurement. Other food applications of MICROTRAC were discussed by Plantz (17).

Similarly, the SVR may be used for particle size analysis of coal oil mixtures. As shown by Morris and Nelson (18), the SVR gives data comparable to the 4-liter aqueous measurement system. The data indicate that analysis in toluene following suspension of coal powder in #6 fuel oil has a negligible effect on the particle size distribution and that suspension in toluene is a viable means of particle size measurement of coal oil mixtures.

Attempts to measure pharmaceutical materials present the need for special handling because of their solubility in both hydrophilic and lipophilic solvents. By performing several sequential analyses on the same sample aliquot, MICROTRAC data indicated definite tendencies of pharmaceuticals to dissolve in many solvents (unpublished data). In cases like these, where aqueous or solvent suspension is not practical, samples may be measured by transportation across the laser beams in a stream of air. With the dry powder sampler (Figure 12), particulate material is loaded into the hopper, from which it is fed onto a vibrating chute. As the sample drops from the chute, it is swept to the sample compartment by means of an air stream, which propels the sample across the path of the laser beam and into a vacuum collector for later disposal. This system provides sufficient energy for the disruption of agglomerates, while not affecting particle integrity. For materials that are extremely fragile, for example, glass spheres that float in most liquids, air flow is reduced to prevent breakage.

Table 2 shows the effect of micronized grinding of USP-grade steroid ester. The data demonstrate that initial crystal size affects final particle

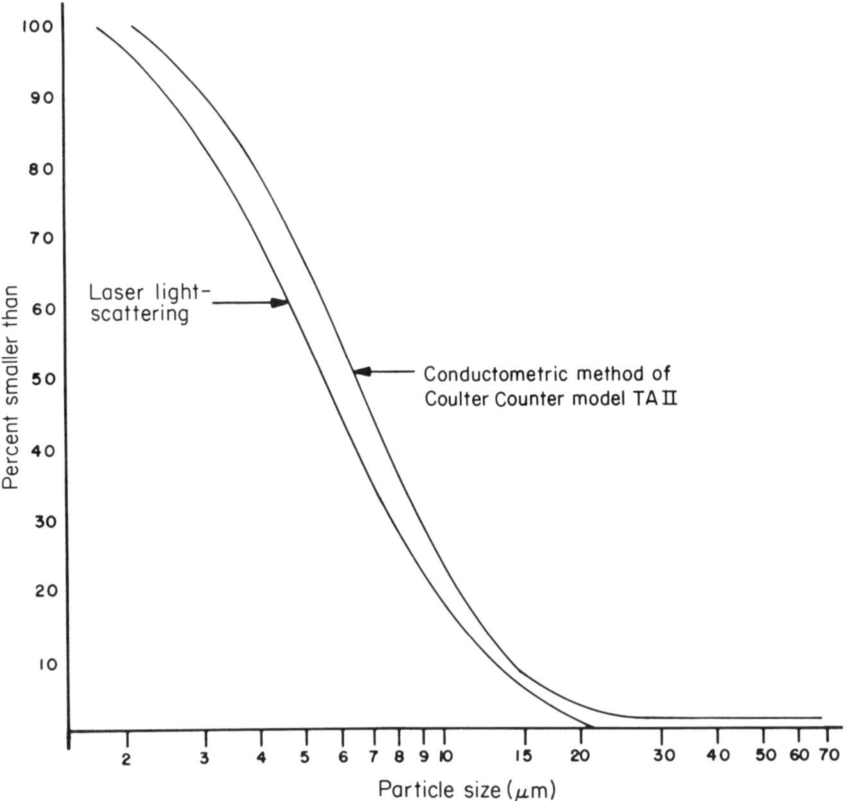

Figure 11. Particle size measurement of chocolate solids obtained by MICROTRAC and Coulter Counter. Courtesy of Instrument Society of America copyright 1979.

size after grinding, which reflects upon the bioavailability of the drug. Similarly, identical grinding of hormone crystals and the ester analogue results in different particle size distributions. By combining drug analogue information and particle size data, one may selectively control bioavailability.

The dry powder sampler is also useful in determining the particle size distribution of cement where determination of surface area as well as percent finer than 44 μm is important to product quality (19). Typically, Blaine (air permeability surface area) and Wagner (sedimentation) particle size determination methods are employed; but, because of the long measuring times necessary, operator dependence, and the requirement for two methods, process control is not timely. Table 3 shows that the MICRO-

Figure 12. Dry powder sampler installed on a standard range MICROTRAC particle size analyzer. The hopper is a funnel-shaped container from which the sample is delivered to the stainless steel chute (30° angle in photo), which delivers the sample to the small black block. The Sample is carried through the "S" shaped pipe to the sample compartment. The vacuum collector (cleaner), which is not shown, is located at the rear of the instrument.

TRAC analyzer employing the dry powder sampler is useful in replacing both methods by virtue of the easily correlatable data.

As with all methods, light scattering has limitations to its applications. As discussed in Chapter 5, concentrations of particulate less than 4×10^{-4} gm/100 ml are not accurately measurable, since the particulates at this concentration are considered to be at contaminating levels. However, concentrations normally found in most commercial processes exceed this value.

Software programming can afford another limitation if appropriate signal collection and algorithms are not employed. In such cases, anomolous distributions may result when fine particulate is over-reported and coarse particulate is under-reported. In addition, some light scattering instru-

Table 2. Comparison of the Particle Size Distributions of a USP-Grade Steroid Hormone Ground by Two Methods[a]

Micron	Normally Ground Steroid Hormone Cumulative Percent Finer	Finely Ground Steroid Hormone Cumulative Percent Finer
176	100	100
125	100	100
88	100	100
62	100	100
44	98.4	100
31	94.6	100
22	88.5	100
16	81.5	100
11	69.7	100
7.8	46.9	94.2
5.5	28	88.3
3.9	14.9	60.8
2.8	5.9	29.7
CS =	0.890	1.74
MV =	11.29	3.97
90% =	24.2	6.18
50% =	8.26	3.52
10% =	3.31	2.20

[a] Originally published in *Drug & Cosmetic Industry*, September 1982.

ments measure an average particle size and then perform curve fitting to provide a distribution. This may cause difficulties in correlating to other light scattering methods or other particle size methods in general. MICROTRAC hardware and software have been evaluated and designed to exclude such problems.

Resolution limitations are not easily overcome by any particle size method, but at the same time, are not of major concern for most operators and applications. Since light scattering techniques require a "cloud" of particles, the resultant scattered light is a combination of many individual diffraction patterns. By means of optical masks described earlier, it is possible to deconvolute the pattern, but some overlap of channel (size) edges exists. This effect is manifested by a slight spreading of the reported distribution. Thus, a monodisperse distribution is usually reported in three

Table 3. Comparison of MICROTRAC Data to Sieve and Sedimentation Data for Three Cement Samples[a]

	Surface Area (m^2/g)			Percent Passing 325-Mesh		
	Sample 1	Sample 2	Sample 3	Sample 1	Sample 2	Sample 3
MICROTRAC[b]	2020	1950	2600	84.4	85.3	94.8
Sieve	—	—	—	91.7	90.3	96.5
Wagner	2010	1850	2220	—	—	—

[a] Reprinted by permission of Cahner Exposition Group.
[b] Specific surface converted to m^2/g using 3.15 g/ml as specific gravity of cement.

channels, where the center channel is the peak containing 25–45% of the volume distribution. On either side of the mode, information is also given.

In practical terms, this means that 5.5-μm monodisperse particulates will show a small but limited response in the 7.8- and 3.9-μm channel. However, consideration should be given to the philosophical idea that absolute resolution may not be achievable technically because of limitations imposed by present day technological and theoretical concepts. Although light scattering manifests such a limitation, application of the technique has not been hampered, as described in Section 6 concerning correlation studies.

4.2. Large Particle Analyzer

An extension of the particle size range of the standard range MICRO-TRAC analyzer to include 44–1000 μm is accommodated also by employing the concept of Fraunhofer diffraction; however, the method by which scattered light flux is resolved and collected is different. Particles larger than those in the range of the standard range instrument scatter light at smaller angles (which are difficult to resolve from the main laser beam). Therefore, an expanded laser beam is used in conjunction with a solid state array (a series of interconnected, light-sensitive silicon diodes). This optical arrangement produces a larger size main beam, capable of being focused to a smaller size while increasing the resolution of scattered light flux near the main axis. The solid state array is employed to collect the highly resolved scattered light. Figure 13 illustrates schematically the large particle analyzer system.

This large particle size analyzer is useful in applications where sieves are generally employed, such as with sand, polyvinylchloride, manganese dioxide aggregates, and mica flakes. As the data in Figure 14 show, particle size measurement, as a part of incoming quality control of highly irregularly-shaped metallic zinc, can be performed on a routine basis and correlated to existing sieve data.

4.3. Small Particle Analyzer

The extension of particle size measurements to particulate material smaller than 2 μm requires special optical and theoretical considerations. As described previously, scattering from particles having sizes at or near the wavelength of the incident light must be treated by Mie theory, which permits measurements to approximately 0.5 μm. Particles from 0.5 to 0.12 μm are measured by employing the 90° scatter theory. The two the-

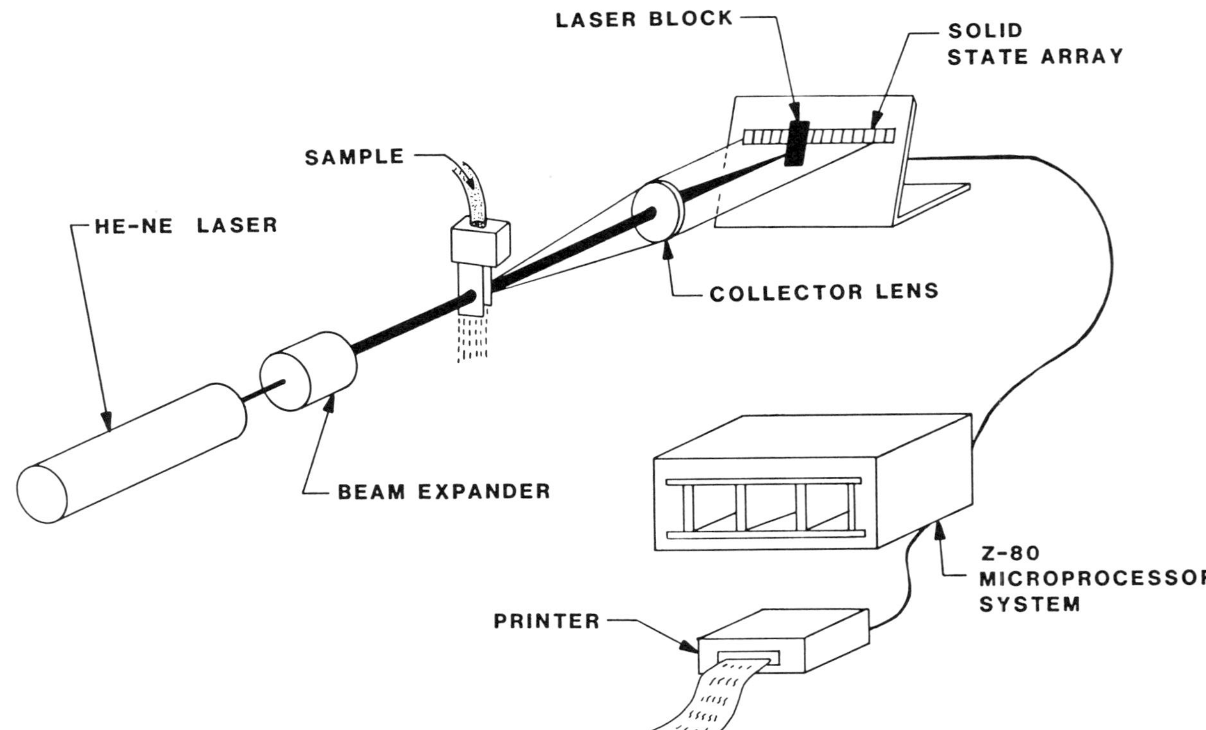

Figure 13. Schematic of the large particle size analyzer optical system. Differences from the standard optical system include the use of a solid state array and beam expander. Both optical systems employ a laser block. Reprinted by permission of Cahner Exposition Group.

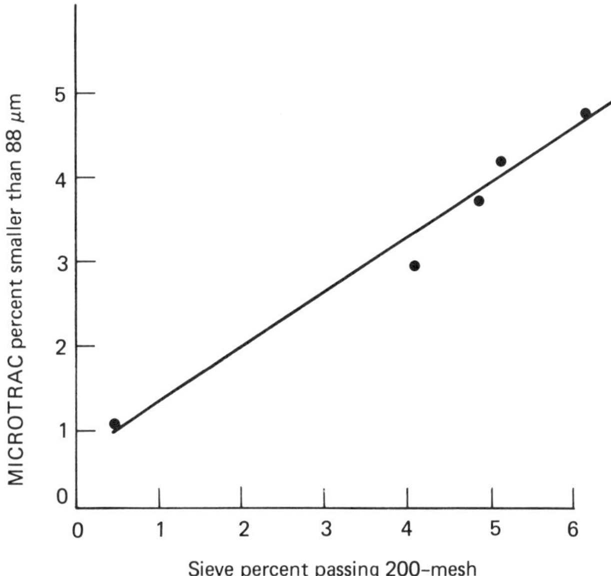

Figure 14.   Comparison of sieve, 200 mesh (74 μm), to MICROTRAC 88-μm channel for zinc metal (the 74-μm channel of MICROTRAC was not used because many values approached 0% passing, which prevents proper correlation). Reprinted by permission of Cahner Exposition Group.

ories are combined in a single instrument by employing two light sources, as shown in Figure 15.

Light from a tungsten–halogen source is passed through a series of six bandpass filters separated into three pairs of three wavelengths. Light from one filter is polarized in the vertical plane, while the second is polarized in the horizontal plane. A circular wheel that houses the six filters (3 wavelengths and 2 planes of polarization for each) rotates synchronously to allow the illumination of particles in the sample cell. Light scattered at 90° is collected normal to the axis of the incident light.

To measure particles from 0.5 to 21 μm, the rotating wheel contains an aperture through which the helium–neon laser passes. During this portion of the sequence, the tungsten–halogen lamp filters are not in place, thus allowing forward scattering, according to the Mie and the Fraunhofer diffraction theories, to prevail, separately from the 90° scatter.

Upon completion of the measurement cycle, light flux data collected by the two detector systems are combined by the computer for reporting of the particle size distribution with the channels shown in Table 4.

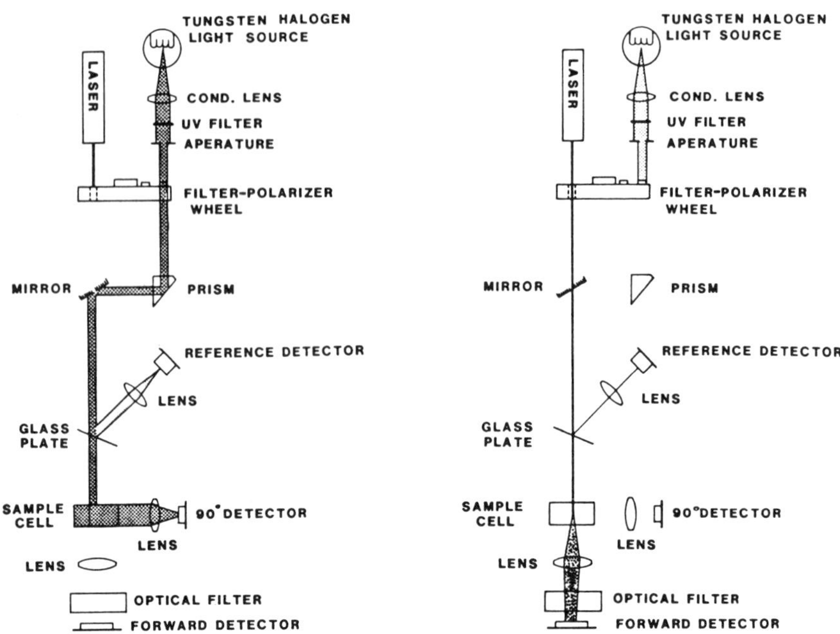

Figure 15. Small particle analyzer, dual optical system. The diagram on the left shows the path of filtered polarized light; the diagram on the right shows the path of the laser source through the optical filter (Compumask). Reprinted by permission of Cahner Exposition Group.

Table 4. Corresponding Channel Centers and Light Sources Employed for Particle Size Measurement over the Range 0.12–21.1 μm[a]

	Light Source			
	Tungsten		Laser	
Wavelength (μm)	0.45 0.60 0.90		0.63	
Channel centers	0.15 0.20 0.30		0.39 0.55 0.80 1.30	
			2.22 3.14 4.44 6.28	
			8.87 12.55 17.75	

[a] Reprinted by permission of Cahner Exposition Group.

194

Table 5.   Comparison of MICROTRAC (SPA) and Mine Safety Appliance (MSA) Median Particle Sizes of Ammonium Perchlorate[a]

Run Number	MICROTRAC Median Particle Size (μm)	MSA Median Particle Size (μm)
1	3.72,3.68	3.90,3.79
2	3.73,3.68	3.95,4.13
3	3.76,3.69	3.90,3.99
4	3.66,3.69	3.94,3.91
5	3.59,3.60	3.74,3.88
6	3.65,3.63	4.01,3.96
7	3.65,3.65	3.79,3.78
8	3.72,3.62	3.85,3.85
9	3.61,3.57	3.74,3.84
10	3.81,3.80	4.03,4.03
Mean	3.67	3.90
Standard deviation	0.067	0.106

[a] Reprinted by permission of E. N. Steger, Atlantic Research Corp.

To evaluate instrument precision, 10 individual runs of fluid energy mill (FEM) grinds of ammonium perchlorate were tested on both the MICROTRAC SPA and the MSA* (centrifugal sedimentation) particle size analyzer (20). Ammonium perchlorate was dispersed in Twitchell Base 8266 and heptane, using an ultrasonic probe at 300 W. As shown in Table 5, the standard deviation of 10 samples measured on MICROTRAC was two-thirds of that obtained from the MSA. Demonstration of instrument repeatability was performed by measuring a single sample of ground ammonium perchlorate relicated 10 times. The relative standard deviation was 0.36% for 4.12 μm particles, which was well within the 0.08 standard deviation expected by the Joint Army Navy NASA Air Force Propulsion Committee for 10 analyses on a single sample.

Zinc oxide (ZnO) was selected as a material requiring sample preparation to demonstrate the ability of the circulating system to maintain dispersion. Approximately 0.5 g of ZnO was dispersed in 10 ml of Tamol L (naphthalene sulfonic acid, Rohm & Haas Co.). Water was added to a final volume of 300 ml; the resulting suspension was blended for 3 min at high speed in a commercial blender. An aliquot of the suspension was then taken for measurement by the 4-liter standard aqueous circulating

* MSA refers to a centrifugal sedimentation method of Mine Safety Appliances that is no longer commercially available, but is still used by some.

system. Figure 16 shows nearly identical particle size distributions of three runs with computed mean and standard deviations of the 10th, 50th, and 90th percentiles of 0.34 (0.01 μm), 0.82 (0.02 μm), and 1.89 (0.05 μm). In the laboratory, metal oxides reagglomerated while the suspensions were quiescent; they required agitation to maintain dispersion. The repeatability of the ZnO measurements shows that the aqueous circulating system is effective in providing the degree of agitation necessary to maintain a fine particle dispersion.

One of the important applications of particle size measuring instruments is to monitor particulate physical changes during grinding operations. Figure 17 is a graphical representation of particle volume mean diameter as a function of relative grinding time for an aqueous dispersion of blue pigment. During initial grinding, the particles showed a typical rapid decrease in mean diameter, followed by a less rapid decrease in mean size. The relationship between the two parameters shows the SPA has sufficient sensitivity to monitor the particle size changes during grinding. This capability is supported by Figure 18, which presents the change in size distribution that occurred when boron was ball milled for 6 hours (20).

Similarly, the ability of the SPA to detect changes in particle agglomeration of ammonium perchlorate when they were exposed to moderately humid conditions was tested (20). As shown in Figure 19, exposure of the chemical to 50% relative humidity for 1 month resulted in a medium particle size change of 44%.

Special conditions are often employed during processing which, because of carrier or medium immiscibility, prohibit the use of aqueous sample preparation. Materials that are manufactured in an organic chemical environment require the use of the same or similar solvent to maintain particulate dispersion. Many solvents exhibit high volatility and flammability, or damaging effects to the 4-liter aqueous circulating system. To avoid such difficulties, a small circulating system manufactured from materials not reactive to most organic chemicals and solvents permits measurement of particles in solvents, such as tetrahydrofuran, dimethyl formamide, cyclohexane, toluene, and alcohols. The small volume circulating system is designed to function similarly to that of the 4-liter system, except for a reduction in liquid to approximately 225 ml. By virtue of this reduced volume requirement, quantities less than 1 g are necessary, while providing the efficient use of expensive solvents. Solvent compatibility of the small circulating system is achieved by a complete Teflon–glass–metal system. Where expensive or highly active chemical reagents are required for dispersion of sample materials, this small volume circulating system is useful as an accessory.

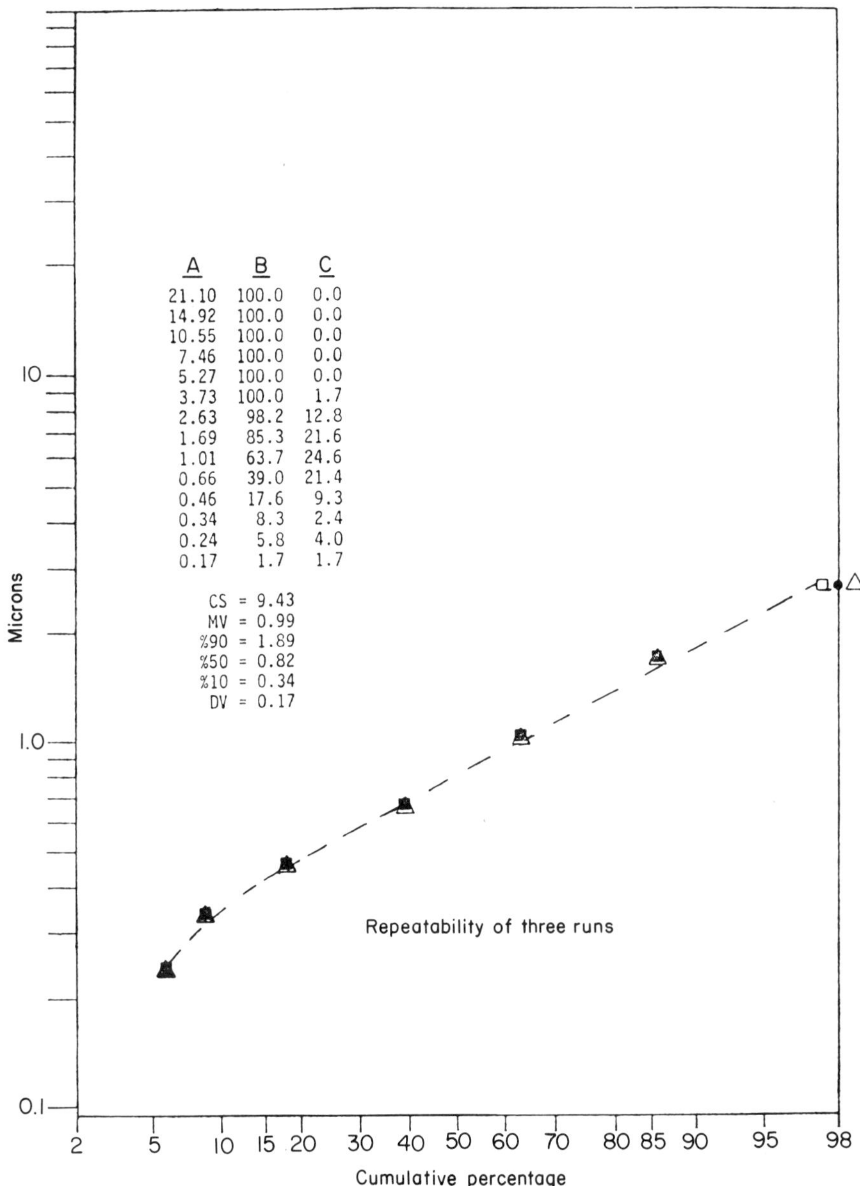

Figure 16. Three sequential measurements of zinc oxide (preparation time—3 min, measuring time—30 sec) demonstrate the ability of the circulation system to maintain the dispersion of a prepared sample. Column A is particle size channel edges, Column B is cumulative percent finer, and Column C is the percent in each channel. Data are the results from one measurement. Reprinted by permission of Cahner Exposition Group.

197

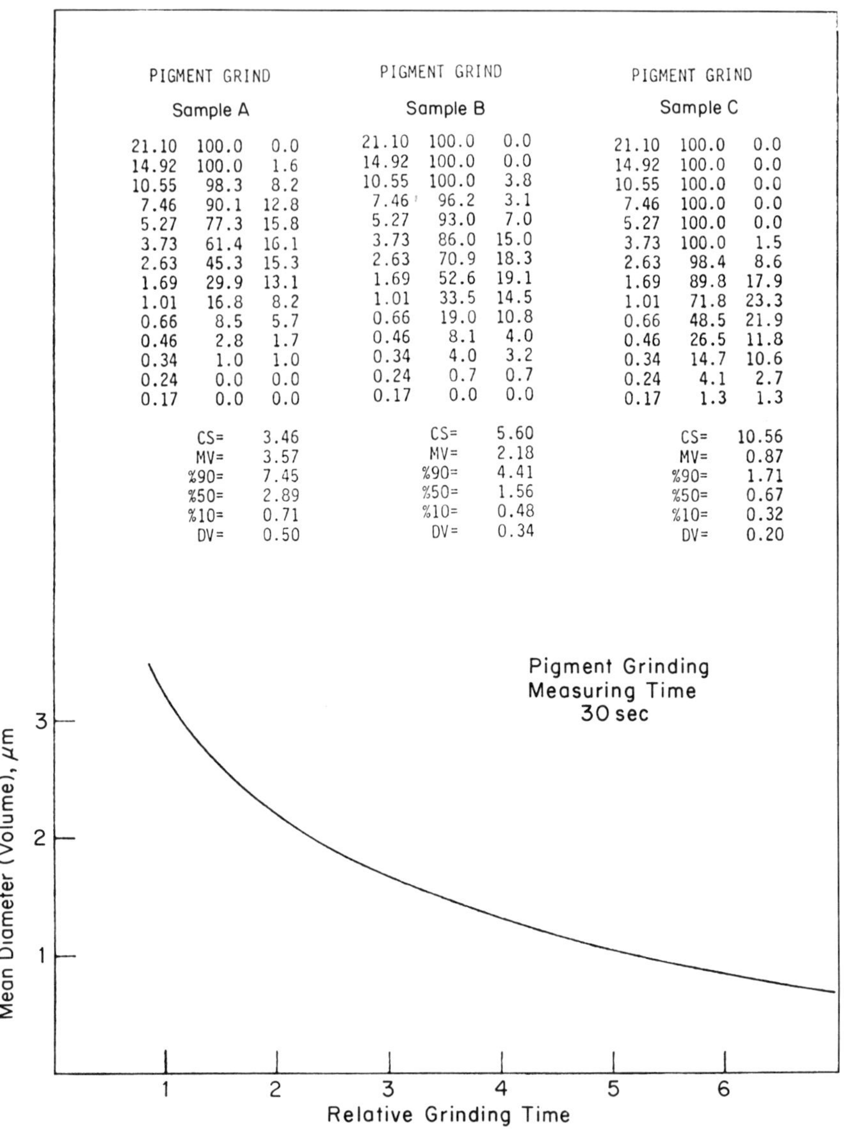

Figure 17. Grinding curve of blue pigment. Reprinted by permission of Cahner Exposition Group. See Figure 6 for an explanation of the data printout.

198

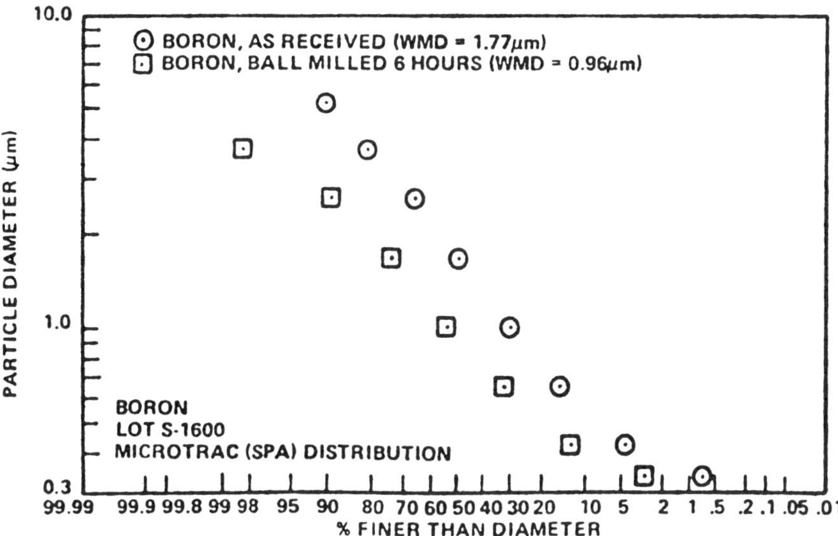

Figure 18.   Effect of 6 h of ball milling on boron, as measured by MICROTRAC SPA. Printed by permission of Atlantic Research Corp. and E. Steger (20).

This system has been applied to the particle size analysis of toner/ Isopar suspensions. Toners for photocopy machines are available as dry powders or liquid slurries in which a lipophilic solvent is used to maintain flowability and dispersion of toner particles. The lipophilic solvent is extremely nonpolar; Isopar, a high boiling point, paraffinic solvent, is effective as a dispersing agent for the toner. In contrast to acetone, Isopar does not damage the integrity of toner particles.

Several types of alumina and synthetic rubber particles are imbedded in a polystyrene matrix to impart various degrees of flexibility and heat resistance. The final characteristics of the plastic are a direct result of the concentration and particle size of rubber. One method of measuring the particle size of such latices is by suspension in solvents such as dimethyl formamide or *trans*-cinnamaldehyde, which do not significantly swell the particles prior to measurement. Hall et al. (21) evaluated the use of the MICROTRAC SPA for particle size measurement of the rubber phase of high-impact polystyrene for purposes of size ranking. Prior to employing a light scattering particle size technique, characterization was performed using microscopic techniques on the rubber sample dispersed in *trans*-cinnamaldehyde. From the study, measurement of the sample dispersed in methyl ethyl ketone by MICROTRAC permitted size ranking

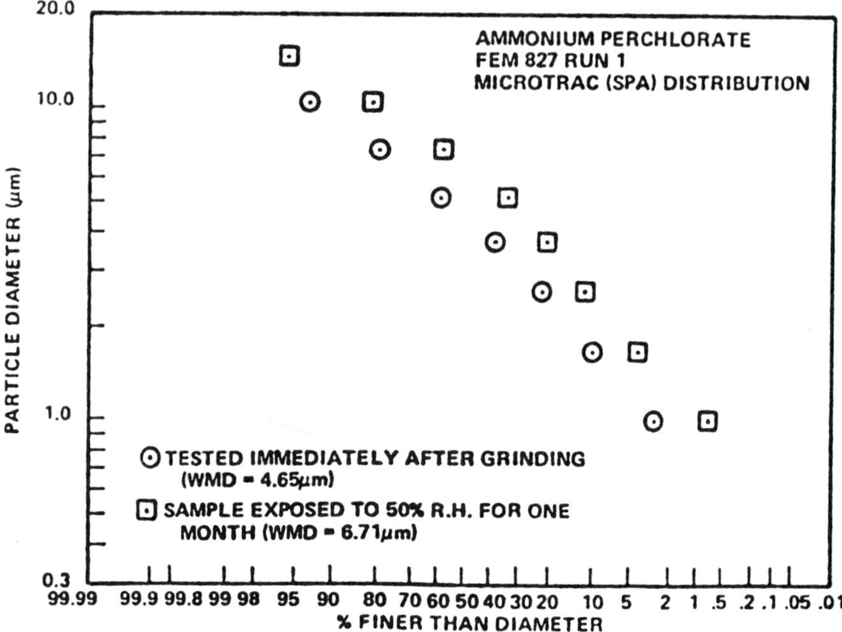

Figure 19. Particle size distribution changes resulting from exposure of ammonium perchlorate to 50% humidity for 1 month. Reprinted by permission of Atlantic Research Corp. and E. Steger (20).

as well as yielded close agreement with microscopic particle size determination.

Magnetic iron oxide is used in the manufacture of recorder tapes, computer floppy disks, and integrated circuits. In the case of an integrated circuit, particle dispersion is of primary importance, since the particles act as gates during the processing of digital data, each particle acting as a single gate. Particle size measurements are used to determine the state of the dispersion, based on size and breadth of the distribution. Methods are available to disperse the iron, but to maintain disorientation of the particles, tetrahydrofuran must be used as a suspending agent. Further, tetrahydrofuran must be used as a diluent for particle size measurements, thus necessitating the use of the solvent resistant MICROTRAC SPA, small volume circulating system.

5. ON-LINE PARTICLE SIZE MEASUREMENTS

The United States consumes vast quantities of energy in comminuation processes such as ore grinding and cement manufacture. Systems management of power can be quite useful in optimizing energy consumption and reducing costs, and on-line process control of particle size can greatly assist conservation by preventing over-grinding. In the past, little attention has been given to such process control because of the need for appropriate particle size sensors. MICROTRAC monitors, an on-line version of the laboratory instrument, can be used to fill the gap between energy consumption and grinding processes. Herbst et al. (22), described the evaluation and use of MICROTRAC characteristics of precision, accuracy, and response time to on-stream measurement of cyclone overflow products. Frock and Challis (23) reported on the reliability of on-line particle size measurements.

Hart (24) described the control of a cyclone classifier using an on-line MICROTRAC monitor and a specialized sample system. The control loop was set to maintain different size distributions of glass-bead feed material. In stability trials, product particle size distributions were maintained within 2% at a set point of 13% passing 11 μm. The system was then stressed with two feed materials of increasing fineness. Cyclone speed varied in response to the particle size measurements to permit compensation of the particle size distribution. Complete correction of the particle distribution was established within three instrument measuring periods (6–9 min), thus showing the efficacy of the control strategy.

A further example of the efficacy of on-line particle size measurement as applied to energy conservation is the cement industry. The particle size of cement can dramatically influence product quality in that cracking and/or fatigue can occur if cement is too fine. If the cement is too coarse, undesirably long curing results. The importance of product quality cannot be overstated, but economic justification of the measurement is also quite important. The finish mill in cement manufacture is used to complete the grinding of the clinker–gypsum mixture, while a separator returns oversized particles to the mill. It has been estimated that, in a typical finish mill, as much as 10% overgrinding can occur to assure sufficient fineness. To reduce costs, on-line particle size control of the finish mill is required.

In addition to the above application of on-line particle size instruments, a particular requirement of automatic control of continuous crystallizers to maintain crystal size distribution has remained a longstanding problem. As reported by Rovang and Randolph et al. (25), the MICROTRAC monitor has been successfully used for continuous crystallizer control and monitoring.

6. CORRELATION AMONG PARTICLE SIZE METHODS

With increased costs of energy, the desire for optimization of grinding processes has become a necessity. In addition, rapid particle size measurement for process and quality control is a requirement to prevent "out-of-spec" products. To permit particle size measurements, many methods, such as electronic zone sensing, sedimentation, sieving, and laser light scattering are available. A common situation arises when the methods described above do not agree with the "absolute" size, which is usually based on a history of results by a single user of one particular method. However, since each method has evolved from individual theoretical concepts and since noise and bias are introduced by the use of manufactured components, absolute agreement among various particle size methods is not yet possible and, therefore, should not be expected.

Various sedimentation methods are widely employed to measure submicron particles. One such method is the Sedigraph 5000, manufactured by Micromeritics Corp., Athens, Georgia. Figure 20 shows a log-probability graph of SPA data for aluminum oxide prepared by suspending it in 0.2% sodium metaphosphate, followed by a 5-min treatment with an ultrasonic probe at 300 W. SPA measuring time for the sample was 30 sec. As the graph shows, sedimentation (using the Sedigraph) reports the particle size as being smaller than that given by MICROTRAC SPA. A contributing factor to this difference is that in the actual practice of particle size measurement, it is well-known that implementation of different particle measuring theories results in different reported data. In addition, implementation of the theories, detector systems, and suspension stability will affect reported data.

Design features of the SPA sampling system produce random particle orientation by incorporating a turbulent mixing system, which causes particle tumbling and maintains particle dispersion. The circulating system prevents sample segregation (which can occur to mixtures of submicron and larger sized particles by virtue of the opposing effects of Brownian movement and gravitation). Thus, segregation effects that may, in part, account for the discrepancy between sedimentation and the SPA are avoided in the SPA measurement. Another effect that generally causes sedimentation devices to report size as being smaller is particle shape. Because of hydrodynamic effects, nonspherical, plate-like particles tend to settle more slowly than spherical ones. The effect of the retarded settling is to bias larger plate-like particles to indicate sizes smaller than equivalent spheres.

In contrast to the above sedimentation data compared to MICROTRAC, centrifugal Mine Safety Appliances measurements (20) tend to report particle sizes that are slightly larger than MICROTRAC SPA. The comparison between MICROTRAC and Mine Safety Appliances is shown in Figure 21, where good correlation is evident between median diameters of 2.5 and 7.0 μm. As median diameter increases above 7 μm, the de-

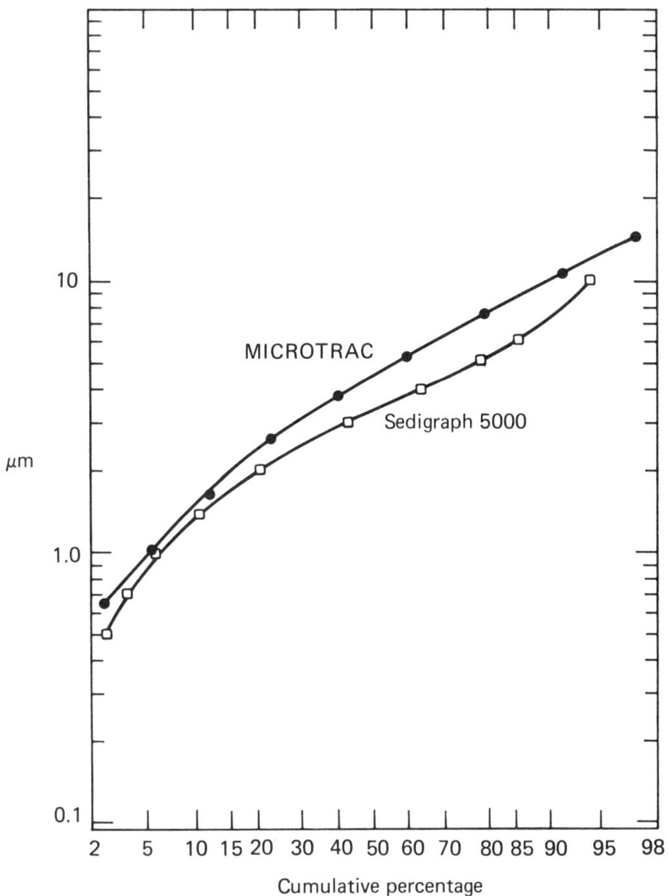

Figure 20. Comparison of MICROTRAC SPA and Micrometrics Sedigraph 5000 for aluminum oxides (preparation time—2.5 min, run time—30 sec). Reprinted by permission of Cahner Exposition Group.

viation from a one-to-one correlation increases, because of the presence of particles too large for measurement by the SPA. (This situation was resolved recently by extending the SPA measurement range to 44 μm.)

To overcome differences in reported particle size and methods, it is important to advance the concept of method correlation, as discussed by Frock and Plantz (26), where, although methods do not agree, particle size changes in a process can be observed by all the methods. In such cases, where two methods report a particle size distribution as coarser or finer, a mathematical relationship can be developed (experience has

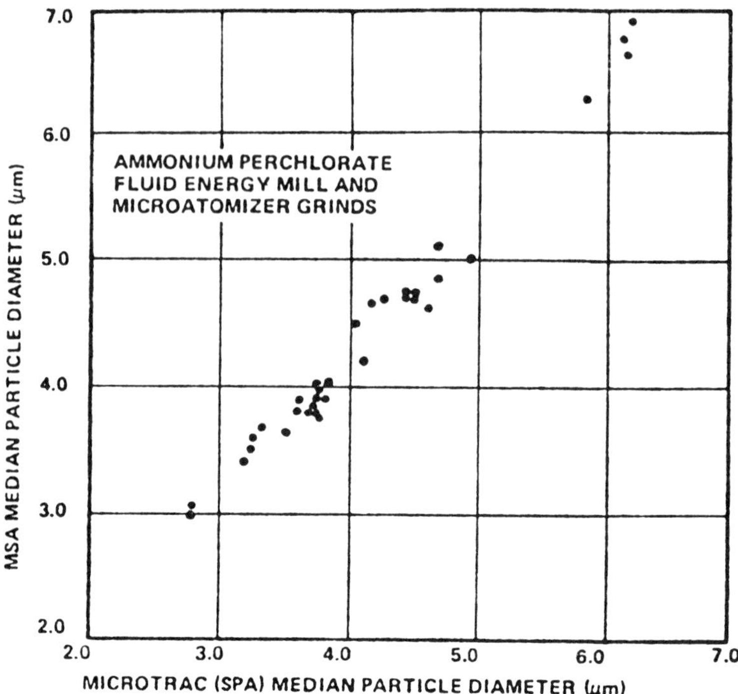

Figure 21.   Comparison of MICROTRAC SPA and MSA sedimentation data from ground ammonium perchlorate products. Reprinted by permission of Atlantic Research Corp. and E. Steger (20).

shown that correlation among three or more methods is usually neither practical nor necessary). A plot of mean diameter, percent passing a given size, or median (50th percentile) diameter of one method is plotted on linear graph paper against a similar value obtained from another method. Examples of such graphical plots are shown in Figures 22–24. In all cases, the graphs show linear plots. The statistical method of least-squares analysis provided the given equation for the best fit of a straight line drawn through the experimentally obtained points. Using such a line or equation, the operator of a new particle size method can quickly and easily relate the data to those from the previous procedure. For example, assume that the particle size specification of alumina has been set at the percent retained on a 44-μm (325-mesh) sieve. According to the correlation, 50% retained in the sieve corresponds to 60% ''greater than,'' as reported by MICROTRAC (Figure 23). Thus, if the specification required at least 50% retained on a 44-μm sieve, the MICROTRAC specification would by 60%

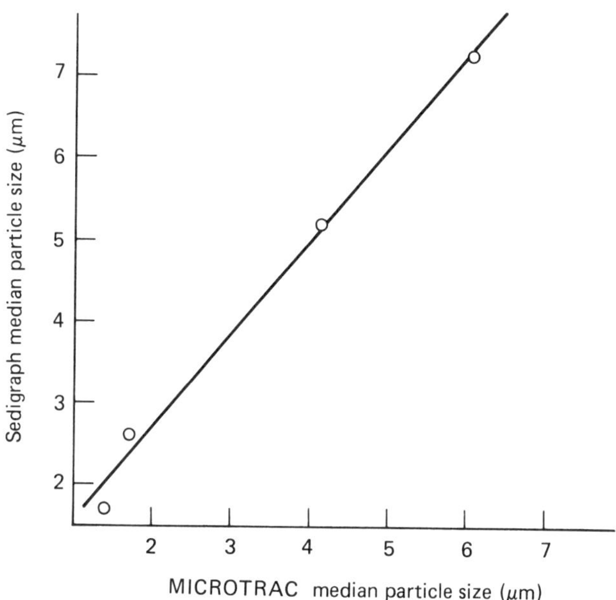

Figure 22. Correlation curve for tungsten powder, MICROTRAC SPA compared to Micromeritics Sedigraph 5000 (correlation equation: $S = 1.16M + 0.383$). Reprinted by permission of Cahner Exposition Group.

greater than 44 μm. In this manner, conversion to a new particle size method is easily accommodated by the quality control laboratory.

7. CONCLUSIONS

This chapter has attempted to demonstrate the technology, usefulness, and application of laser light scattering, as implemented in the line of MICROTRAC particle size analyzers, to a wide variety of industrial processes, as shown in Table 6. With increased energy and labor costs, the requirement and application of particle size analysis has become an important integral part of quality and process control operations. Through efficient grinding operations monitored by real-time particle size analyzers, energy consumption can be reduced while output is increased. Emulsion characteristics and stability can be studied to increase shelf-life, and the influence of particle size on bioavailability and tableting can be determined (see Chapter 2). In addition, on-line particle size monitors permit unattended measurement and immediate and timely manual or automatic

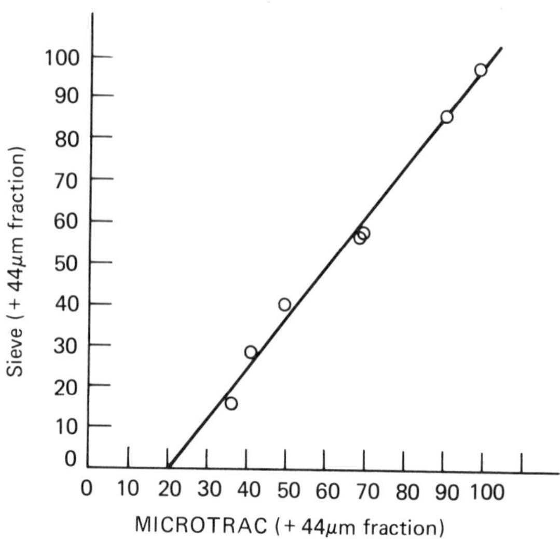

Figure 23. Correlation curve for alumina, MICROTRAC standard range analyzer (1.9–176 µm) compared to sieve measurement (correlation equation: $S = 1.25M - 25.4$). Reprinted by permission of Cahner Exposition Group.

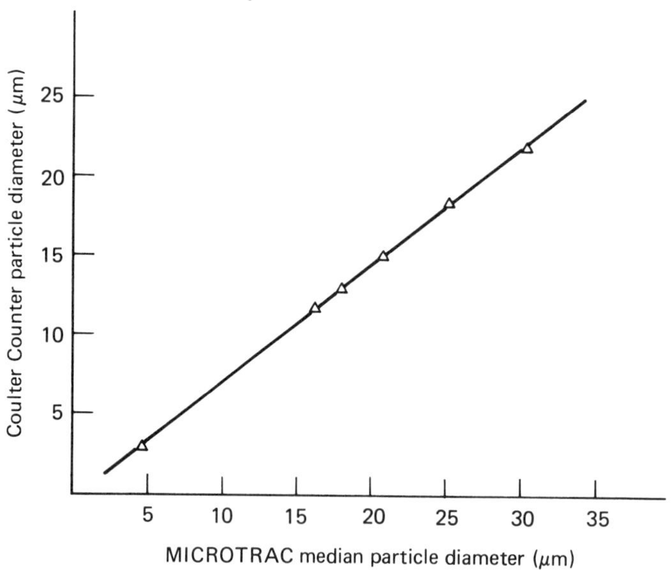

Figure 24. Correlation curve for alumina, compared to MICROTRAC standard range analyzer and Coulter Counter data (correlation equation: $C = 0.758M - 0.337$). Reprinted by permission of Cahner Exposition Group.

206

**Table 6. Some Materials Measured and Processes Monitored by the MICROTRAC
Light Scattering Technique**

Adhesives	Graphite
Alumina	Graphite production
Aluminum powders	Iron ore
Analgesics	Jet engine development
Apple juice	Lead oxide
Battery production	Manganese dioxide
Calcium carbonate	Mineral processing
Carbon powders	Oil/water emulsions
Catalysts	Oil-well drilling
Cement	Pesticides
Cement grinding	Pharmaceuticals for tabletting
Ceramics	Phosphate processing
Chocolate	Pigment production
Chromium dioxide	Pigments
Clarifier testing	Polyvinyl chloride
Clays	Potassium chloride, saturated
Coal	Resins
Coal and oil mixtures	Rocket propellants
Coatings	Rolling mill emulsions
Cocoa	Shale oil
Coffee grinding	Silica
Combustion control	Silicon carbide
Copper powder	Spray drying
Corn flour	Synfuels
Cosmetic manufacturers (wet and dry)	Taconite processing
Crystallization	Talc
Drilling muds	Tantalum
Explosives	Tomato juice
Filter testing	Toners
Flocculation studies	Tar sands
Fly ash	Tungsten powders
Garnet	Wheat flour
Geothermal development	Zirconium milling
Glass beads	

feedback control. These and other applications over the particle size range
of 0.1–1000 μm are presently possible in the laboratory or for on-line
usage. The same technology of laser light scattering thus facilitates the
transfer of laboratory particle size measurements to the process location
for a broad spectrum of powders and slurries.

REFERENCES

1. E. C. Muly, H. N. Frock, and E. L. Weiss, *J. Powder Bulk Solids Technol.*, **2**, 3 (1978)

2. G. Mie, *Ann. Phys.*, **25**(3), 377 (1908).

3. A. L. Wertheimer and W. L. Wilcock, *Appl. Opt.*, **15**, 1616 (1976).

4. E. L. Weiss and H. N. Frock, *Powder Technol.*, **14**, 287 (1976).

5. E. C. Muly, H. N. Frock, and E. L. Weiss, "The Application of Fourier Imagery Systems to Fine Particles," Seventh Annual Fine Particle Conference, Philadelphia, Pa., 1975.

6. E. C. Muly, H. N. Frock, and H. M. Rohr, "Extended Range Instrumentation for Large Particle Analysis," Powders and Bulk Solids Conference, Chicago, Ill., May 14, 1980.

7. E. C. Muly and H. N. Frock, "Submicron Particle Size Analysis Using Light Scattering," Fine Particles Society Fall Meeting, University of Maryland, College Park, Md., 1980.

8. W. L. Anderson and R. E. Berssner, *Appl. Opt.*, **10**, 1503 (1971).

9. H. C. Van de Hulst, *Light Scattering by Small Particles*, Wiley, New York, 1957, pp. 174–180.

10. T. E. Morris and H. Bell, *Mod. Dev. Powder Metall.*, **12**, 31 (1980).

11. E. C. Muly and H. N. Frock, *J. Opt. Eng.*, **14**, 861 (1980).

12. P. E. Plantz and C. M. T. Lowe, *Drug Cosmet. Ind.*, **131**(9), 30 (1982).

13. A. B. Ward, D. Wetzel, and J. Vetter, "Rapid Particle Size Distribution Analysis in Milling and Baking," 62nd Annual Meeting of the American Association of Cereal Chemists, San Francisco, Ca., October 22–23, 1977.

14. H. N. Frock and R. L. Walton, *Am. Ceramic Soc. Bull,* **59**, 650 (1980).

15. R. L. Haverland and L. R. Cooper, Proceedings of the Arizona Section, American Water Resources Association, Nevada Academy of Science, Volume 2, 1981, p. 207.

16. E. B. Kinter and S. Diamond, Nat. Acad. Sci., National Research Council Publication 566, 1958, pp 318–333.

17. P. E. Plantz, "Particle Size Measurements of Food Products Using Laser Light Scattering," Instrument Society of America, National Conference and Exhibit, Chicago, Ill., October 22–25, 1979.

18. T. E. Morris and R. Nelson, "Particle Size Determination of Coal in Coal Oil Mixtures," 1981 Symposium of Instrumentation and Control for Fossil Energy Processes, San Francisco, Ca.

19. P. E. Plantz, "The Use of Particle Size Measurement to Minimize Energy Consumption and to Maximize Plant Through-Put," Powder and Bulk Solids Conference, Chicago, Ill., May 21, 1982.

20. E. H. Steger, "Evaluation of the MICROTRAC (SPA) for Particle Size Analysis of Superfine Materials, JANNAF Propellant Characterization Subcommittee Meeting, USAFA, Colorado Springs, Co., August 18, 1981.

21. R. A. Hall, R. O. Hites, and P. E. Plantz, *J. Appl. Polym. Sci.*, **27**, 2885 (1982).

22. J. A. HERBST, D. J. Kenneberg, and K. Rajamani, Proceedings of International Symposium on In-Stream Measurement of Particulate Solid Properties, Volume 1, Bergen, Norway, August 1978.

23. H. N. Frock and W. Challis, "Reliability of On-Line Particle Size Measurements," Powder and Bulk Solids Conference, Chicago, Ill., May 19, 1981.

24. W. H. Hart, "Description and Performance of a Dry Powde Pilot Scale Process Control System," Powder and Bulk Solids Conference, Chicago, Ill., May 18, 1981.

25. R. Rovang and A. Randolph, "On-Line Particle Size Analysis in the Fines Loop of a KCl Crystallizer," 71st American Institute of Chemical Engineers Meeting, Miami, Fl., November 1978.

26. H. N. Frock and P. E. Plantz, "Correlation among Particle Sizing Methods," Powder and Bulk Solids Conference, Chicago, Ill., May 21, 1982.

CHAPTER

7

FIELD-FLOW FRACTIONATION
OF PARTICLES

KARIN D. CALDWELL

Department of Chemistry
University of Utah
Salt Lake City, Utah

1. INTRODUCTION

"Field-flow fractionation" (FFF) is the generic name for a class of separation techniques that are ideally suited for fractionation and characterization of particulate as well as soluble samples. Whether in the colloidal size range or above, monodisperse or polydisperse, particles may be characterized as to mass, size, or other properties based on their fractionation pattern.

The fractionation occurs in a narrow, ribbonlike channel, whose dimensions can be selected to suit a particular sample. The field, which is responsible for the selectivity of the procedure, can be chosen to give maximum discrimination between different types of particles in a sample. As a result of this versatility, FFF in one form or another has been applied to the characterization of particles or molecules whose sizes range over 5 orders of magnitude. Samples include charged and uncharged species that are suspended or dissolved in aqueous as well as organic media. Mass, size, charge, or other pertinent sample characteristics are calculated from the elution pattern generated when a particulate sample undergoes field-flow fractionation. Since elution patterns are usually recorded by a nondestructive detection method, fractions can be collected for further chemical, physical, or biological analysis, thus maximizing the information that can be gathered from a given sample.

This method is a one-phase separation technique in which a field, applied perpendicular to a particle-containing flow stream, forces dissimilar particles to partition among flow lines of different velocity. Differences in field interaction lead to differences in migration velocity, which results in chromatographic resolution.

The FFF concept was formulated by Giddings in 1965 (1, 2), and during the years following, much work at our University of Utah laboratory has been devoted to practical implementation of the theory. Of the different fields that were explored (3–7), only sedimentation (6) and flow (7) have had application to particulate samples and warrant discussion in this chapter. Although most applications presented originated in our laboratory, various forms of FFF are currently under study by a number of research groups in the United States and Europe (8–11). Of particular importance in this context is the work by Kirkland, Yau, and coworkers at the DuPont Experimental Station, whose contributions (12–14) have significantly advanced the technology of particle sizing by means of sedimentation FFF. As a result, particle size characterization can now be performed in a few minutes.

2. PRINCIPLES OF OPERATION

2.1. General

Field-flow fractionation is a one-phase chromatography related technique (15), where an externally adjusted field is allowed to interact with a par-

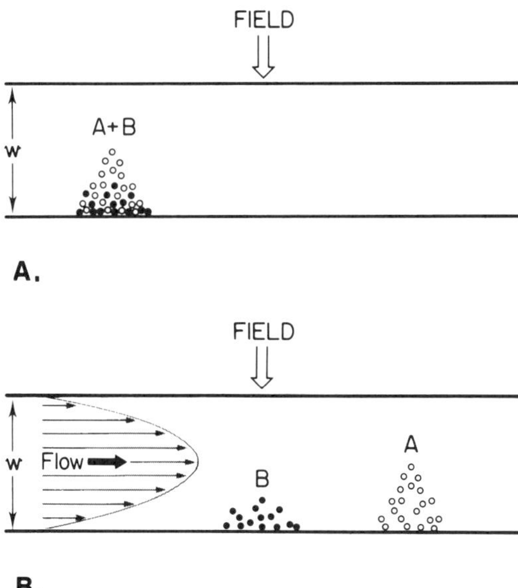

Figure 1. Principle of FFF: (*A*) prior to the onset of flow, components A and B have reached their equilibrium distribution in the channel. (*B*) A laminar flow perpendicular to the field moves the two particle types downstream at different velocities.

ticle suspension under motion in a duct or channel. Because the channel is open and unobstructed, complex geometry-dependent factors, such as particle size or shape, play an insignificant role in normal FFF operation. If the geometry of the duct is known, the elution behavior of a particle influenced by a given field can be directly related to the magnitude of the physical property that made the particle susceptible to the field. Although the principle operates equally well in ducts of cylindrical geometry (8), provided that velocities are sufficiently low to prevent secondary flow, we have preferred to work with channels of rectangular cross section, which are easy to assemble and dismantle. The channel configuration determines the flow profile of the carrier medium as it passes through the duct; the applied field governs the position of the solute zone within this profile (Figure 1) and, as such, the average zone velocity. It becomes evident that chromatographic retention can be predicted from a knowledge of the zone thickness. Conversely, an observed retention is uniquely related to a particular layer thickness that, at a given field, relates to some property of the particle.

2.1.1. The Particle Zone

Immediately following injection of a particulate sample, the channel flow is stopped and the field is applied to the channel. Particles begin to migrate with the field toward positions of lower potential energy, but their migration is stopped by the presence of the channel wall. As a result, an accumulation takes place in the wall region, which is more or less offset by dispersion owing to Brownian motion. At equilibrium, the particle zone extends exponentially from the accumulation wall (16). The equation

$$c(x) = c_0 \cdot e^{-x/l} \tag{1}$$

describes the concentration distribution in the x-direction (this is the direction of higher potential), which is perpendicular to the wall ($x = 0$). Here, c_0 represents the concentration at the wall and l is a layer thickness parameter that equals the ratio of the particle's diffusion coefficient D to its field induced velocity U (2)

$$l = \frac{D}{U} \tag{2}$$

If the field exerts a force F on a particle with friction coefficient f, the velocity U can be replaced by F/f. The Einstein relationship expresses the diffusion coefficient D as kT/f, where k is the Boltzmann constant and T is the temperature. With these substitutions, Equation 2 becomes:

$$l = \frac{kT}{F} \tag{3}$$

For the following discussion of FFF retention, it is convenient to introduce the "reduced layer thickness" λ as the ratio of zonal thickness l to the thickness of the channel, w:

$$\lambda = \frac{l}{w} \tag{4}$$

Equation 1 is only valid when particle dimensions can be neglected in comparison with layer thickness l. When this is no longer the case, the separation is influenced by steric effects. The ultimate extreme arises when the particle radius r is greater than the ideal layer thickness l. This condition, referred to as steric FFF (17), constitutes a limiting condition for all normal FFF operation.

2.1.2. Retention

Once the zone has relaxed into its equilibrium configuration, the channel flow is resumed and the originally superimposed zones of different particle types begin to migrate downstream in the channel. The velocity of each zone is calculated as an average over all particles in the zone:

$$v_{zone} = \frac{\langle v(x) \cdot c(x) \rangle}{\langle c(x) \rangle} \tag{5}$$

To compute this cross-secional average, one needs to know both the functional form of the velocity profile $v(x)$, which is determined by the chosen channel geometry, and the distribution of particles within the channel, which is expressed by Equation 1. In channels whose width (w) is much less than the breadth, the velocity profile is well-described by the "infinite parallel plate" approximation for laminar flow (18):

$$v(x) = 6 \langle v \rangle \left(\frac{x}{w} - \frac{x^2}{w^2} \right) \tag{6}$$

where $\langle v \rangle$ is the average linear velocity between plates with a spacing of w.

In chromatographic theory, the retention factor R (defined as R_f in Chapter 9) is defined as the ratio of sample velocity to the velocity of the carrier (19):

$$R = \frac{v_{zone}}{\langle v \rangle} \tag{7}$$

Since v_{zone} and $\langle v \rangle$ are inversely proportional to the elution times (or elution volumes) for the zone and for a nonretained peak, the ratio R can be written as

$$R = \frac{t^0}{t_e} = \frac{V^0}{V_e} \tag{8}$$

where t^0 and t_e are elution times, and V^0 and V_e are elution volumes for unretained and retained peaks, respectively. Equation 7 can now be used to formulate a relationship between retention R and the layer thickness

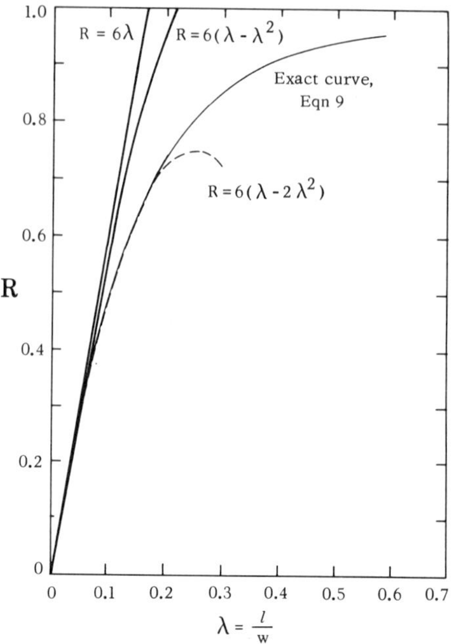

Figure 2.   The exact retention curve for normal FFF in an infinite flat channel, as obtained from Equation 9, and various approximations valid at high retention (low *R* values). Reprinted from reference 50 with permission.

of a particle zone. By introducing the expression for v_{zone} given in Equation 5, and replacing l/w with λ, one obtains (16)

$$R = 6\lambda \left[\coth \frac{1}{2\lambda} - 2\lambda \right] \tag{9}$$

This relationship is illustrated in Figure 2. The bracketed expression rapidly approaches unity as λ decreases, so that in the limit of small λ values, R becomes a linear function of the reduced layer thickness:

$$\lim_{\lambda \to 0} R = 6\lambda \tag{10}$$

As seen in Figure 2, this relationship holds for $R < 0.1$. From Equation 8, it follows that this applies to zones retained beyond 10-column volumes. For the elution interval between 2- and 10-column volumes, a reasonable approximation becomes

$$R \simeq 6\lambda - 12\lambda^2 \tag{11}$$

As noted in Section 2.1.1, deviations from this ideal behavior occur for particles whose dimensions approach the particle layer thickness l. Because of their size, such particles are excluded from the immediate wall region of the channel, since they can approach the wall no closer than one particle radius, r. Equation 9 can be modified to account for this effect (20):

$$R = 6\gamma(\alpha - \alpha^2) + 6\lambda(1 - 2\alpha)\left[\coth\left(\frac{1 - 2\alpha}{2\lambda}\right) - \frac{2\lambda}{1 - 2\alpha}\right] \tag{12}$$

where $\alpha = r/w$ and γ is a velocity-dependent factor whose value is near unity. For large particles, the retention becomes entirely governed by their size through α, rather than by their reduced layer thickness (λ). Thus, for $\alpha \gg \lambda$, R takes the limiting form

$$R = 6\gamma\alpha \tag{13}$$

When such conditions prevail, the separation mechanism is referred to as "steric FFF."

2.1.3. Zone Broadening

Efficient separation of mixtures of components is the result not only of differences in retention, but also of minimal zone broadening incurred during passage through the channel. As in chromatography (19), zone broadening in FFF (2, 16) is discussed in terms of plate height, H, which is defined in terms of the variance, σ^2, generated within the zone per unit of migration distance. For a uniform channel of length L, H becomes

$$H = \frac{d\sigma^2}{dx} = \frac{\sigma^2}{L} \tag{14}$$

The various factors that contribute to the broadening of solute zones in FFF may be summarized as follows (21):

$$H = H_l + H_n + H_p + \Sigma H_i \tag{15}$$

Here H_l accounts for broadening caused by diffusion during the particle's column residency. At normal flow rates, this effect can be neglected because of the longer time scale of particle diffusion. The term H_n denotes the nonequilibrium effects on plate height, which arise from slow mass transport between flow lines of different velocity. Unlike H_l, this term gains in importance with decreasing particle diffusivity. The term is velocity dependent and has the following form

$$H_n = \chi \cdot \frac{w^2 \cdot \langle v \rangle}{D} \tag{16}$$

where w is the channel thickness, D is the particle's diffusion coefficient, and $\langle v \rangle$ is the average flow velocity in the channel. The parameter χ depends on λ; it accounts for the reduction in H_n from field compression of the particle zone. The complete expression for χ is somewhat complex (22), but in the limit of high retention it reduces to

$$\lim_{\lambda \to 0} \chi = 24\lambda^3 \tag{17}$$

This limiting form is accurate to within 10% for samples retained for more than 14-column volumes.

Polydisperse samples broaden as a result of fractionation in the channel and give rise to plate height contribution H_p. In terms of the standard deviation of the particle diameter within the zone, σ_d, the polydispersity contribution depends on the retention parameters R and λ, as well as on the average particle diameter $\bar{d}$ (23):

$$H_p = L\left(\frac{1}{R} \cdot \frac{dR}{d\lambda} \cdot \frac{d\lambda}{d\bar{d}}\right)^2 \cdot \sigma_d^2 \tag{18}$$

The factor $d\lambda/d\bar{d}$ depends on the type of field employed for the separation; it is evaluated below in conjunction with the respective subtechnique.

Nonideality contributions to plate height are accounted for by the composite term $\Sigma\, H_i$ in Equation 15. Among such contributions are the effects of imperfect injection and channel design and spreading in the detector cell. Neglecting H_l, under ideal conditions, H_n is the only term with an explicit velocity dependence; hence, Equation 15 may be reduced to (24)

$$H = C\langle v \rangle + A \tag{19}$$

where C and A are constants independent of velocity. Equation 19 predicts that the plate height will vary linearly with carrier velocity. The slope of the line, C, is seen from Equation 16 to depend on the channel thickness w, the sample diffusion coefficient D, and the nonequilibrium coefficient $\chi(\lambda)$, whose value can be computed from the observed, retention-derived λ:

$$C = \chi(\lambda)\,\frac{w^2}{D} \tag{20}$$

When χ and w are known, D can be evaluated directly from the slope of the plate height curve. The Stokes–Einstein equation relates the diffusivity of a particle to its diameter, if the particle is spherical, or to the diameter of an equivalent sphere with identical friction in a suspension medium with viscosity η:

$$D = \frac{kT}{f} = \frac{kT}{3\pi\eta d} \tag{21}$$

The intercept A of the plate height curve represents the sum of all velocity-independent contributions to H. In the absence of nonideal zone broadening, this intercept reflects the polydispersity of the sample through Equation 18. By systematically studying zone broadening of samples with known particle diameter and polydispersity, one has a powerful diagnostic tool for evaluating channel design and performance. A departure from linearity in the plate height curve, or intercepts in excess of the values predicted by Equation 18 signal nonideal performance by the equipment.

2.2. Specific Subtechniques

Although there is no inherent limit to the types of fields that are potentially useful in FFF, the present discussion concentrates on sedimentation and flow FFF, with their limiting form, steric FFF. In recent years, much effort has been devoted to the evaluation of these techniques. This work has indicated that sedimentation FFF shows particular promise for the rapid and selective analysis of colloidal materials.

The particle size range that is currently amenable to analysis stretches from 1 nm for the flow FFF and 10 nm for the sedimentation analogue to 1 μm, where steric effects become significant. In the fully developed steric regime, particles with diameters up to 100 μm were analyzed. By varying the field strength, one can raise or lower the size range where particles become susceptible to steric effects, as discussed below.

2.2.1. *Sedimentation FFF*

In the presence of a gravitational field, particles are affected by a force
F whose magnitude depends on the mass of the particle as well as on the
gravitational acceleration. If the force field is generated in a centrifuge,
the acceleration is determined by the angular velocity ω and the radius
r' of the rotor. For particles suspended in a condensed medium, the "effective mass," m', represents the total mass m of the particle adjusted
for the buoyancy of the medium (25). Thus,

$$F = m' \cdot \omega^2 r' = m\left(1 - \frac{\rho}{\rho_p}\right)\omega^2 r' \tag{22}$$

where ρ and ρ_p are the densities of solvent and particle, respectively. By
inserting the force expression of Equation 22 into the equations for reduced layer thickness in FFF (Equations 3 and 4), one obtains (23):

$$\lambda_s = \frac{kT}{m(1 - \rho/\rho_p) \cdot \omega^2\, r' \cdot w} = \frac{kT}{m \cdot \Delta\rho/\rho_p \cdot G \cdot w} \tag{23}$$

where $\Delta\rho$ denotes the density difference between particle and suspension
medium, and G replaces $\omega^2 r'$ as the notation for the gravitational acceleration. It is often useful to cast λ in terms of particle diameter d rather
than mass m for a retained particle. For spherical particles, m in Equation
23 can be replaced by the product of density and volume, where the latter
is expressed in terms of d:

$$\lambda_s = \frac{6kT}{d^3 \cdot \pi \cdot \Delta\rho \cdot G \cdot w} \tag{24}$$

For nonspherical particles, d represents the diameter of a sphere whose
volume is equivalent to that of the particle.

In the case of sedimentation FFF, the reduced layer thickness λ is a
function of the chosen experimental parameters T, G, and w, as well as
of the sample characteristics mass or diameter and density. The inverse
relationship between λ and the selected field strength G is illustrated in
Figure 3. Equation 9 was used here to convert experimental retentions
into the corresponding λ values for a set of polydisperse latices with specific diameters. For each particle, λ varies linearly with $1/G$, as predicted
by Equations 23 and 24. Evaluation of particle dimensions from retention
data is somewhat more cumbersome, since λ depends on both particle

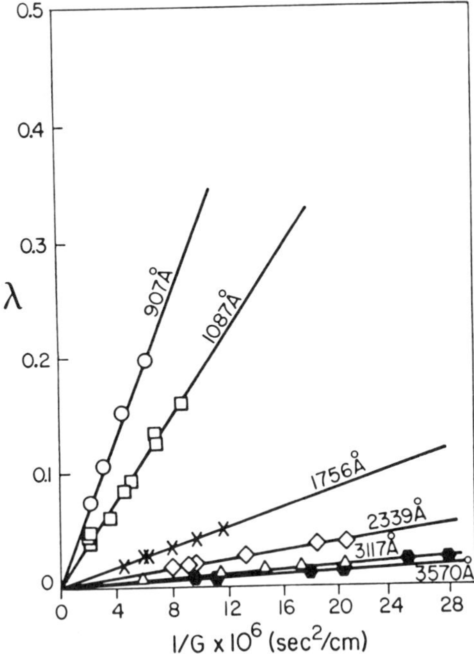

Figure 3. Variations of retention parameter λ with reciprocal field strength 1/G for various particle diameters. The straight lines are obtained from Equation 24. Reprinted from reference 23 with permission. Copyright 1974 American Chemical Society.

mass (or diameter) and density. Where density data are available from outside sources, the mass or diameter is directly determined from a particle's retention at a given field. In the absence of such information, both density and mass can be evaluated for a given sample by systematic observations of retention in carriers of different density. Figure 4 demonstrates how variations in carrier density, and thus variations in the density difference between carrier and particle Δρ, lead to variations in retention (26).

A convenient form for graphic evaluation of particle mass or diameter and density from such measurements is obtained by the inverse of λ_s (Equation 23). This expression is linear with respect to the carrier density ρ:

$$\frac{1}{\lambda_s} = \frac{m\,G\,w}{kT} - \frac{m\,G\,w}{\rho_p kT} \cdot \rho \tag{25}$$

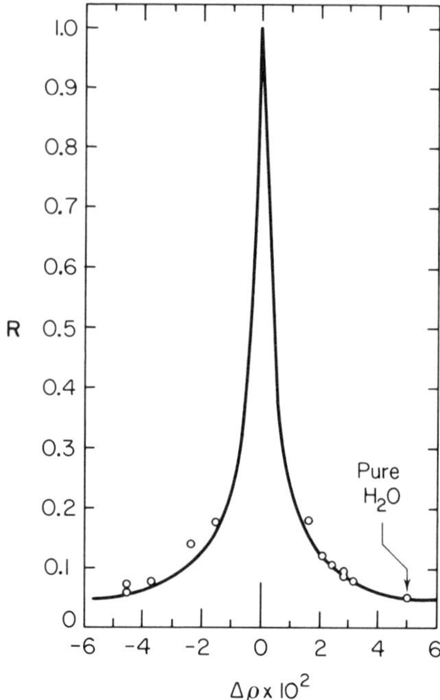

Figure 4. Retention factor R vs. density difference $\Delta\rho$ for 0.312 μm polystyrene beads in various aqueous solutions of sucrose. The field was 89.5 g. The solid lines are obtained from theory. Reprinted from reference 26 with permission. Copyright 1974 American Chemical Society.

By multiplying both sides of the equation by kT/Gw and replacing $kT/Gw\lambda_s$ with the reduced mass, m', Equation 25 is transformed into (27):

$$m' = m - \frac{m}{\rho_p} \cdot \rho \tag{26}$$

An equivalent form can be cast in terms of the particle diameter d:

$$\rho = \rho_p - \frac{1}{0.52360d^3} \cdot m' \tag{27}$$

A plot of the experimentally derived parameter m' against carrier density ρ gives a straight line, whose intercept equals the mass of the particle. Once the mass value is known, the particle density ρ_p is easily evaluated from the slope of the line. Figure 5 illustrates an application of this pro-

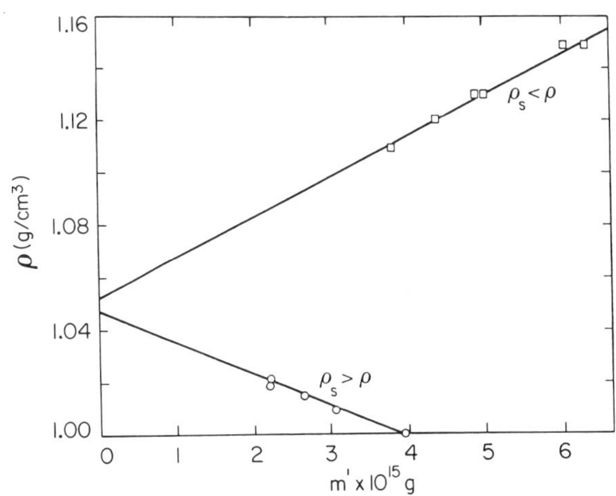

Figure 5. Plot of ρ vs. m' for polystyrene latex beads of nominal diameter 0.481 μm. Reprinted from reference 27, by courtesy of Marcel Dekker, Inc.

cedure to a polystyrene latex sample that was suspended in an aqueous detergent solution. Through additions of sucrose, the density of the suspension medium was readily varied to levels both above and below the density of the particle. A similar approach was used to study a sample of T4D virus (28); however, in this case, cesium chloride was added to increase carrier density.

In the general discussion of retention given above, it was noted that a particle's migration velocity is solely governed by λ, only when its dimensions can be neglected in comparison with the thickness of the solute layer. When this is no longer the case, the presence of steric effects requires the use of Equation 12 for the proper evaluation of particle diameters from retention. This relationship is illustrated graphically in Figure 6, where R is plotted as a function of particle diameter at a given field and density difference (29). The three regions corresponding to normal FFF behavior (negligible steric effects), transition behavior (mixed effects), and steric FFF (idealized layer thickness negligible) are indicated in the diagram. By varying G and $\Delta\rho$, one is free to move the region boundaries toward higher or lower values of d.

A good particle separation process effectively sorts particles of different diameters according to size. If the sorting process is an elution technique, and if zone broadening effects are minimized, the best resolution is obtained when a small difference in diameter effects a large difference in elution volume. By combining Equations 8 and 12, and differ-

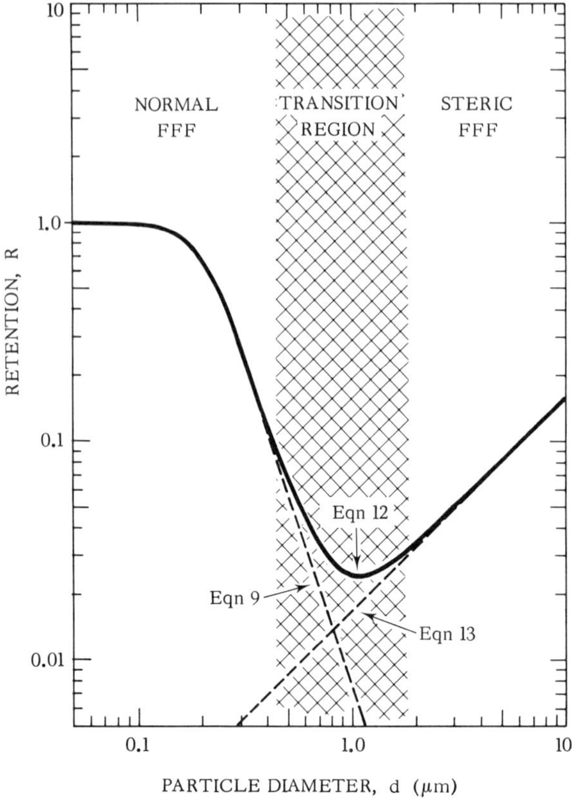

Figure 6. The variation in retention R with particle diameter for sedimentation FFF. The field is 100 g, the density difference is 0.05 g/cm^3 for a channel thickness w of 0.0127 cm. Parameter γ in Equation 12 is assumed to be 0.7.

entiating V_e with respect to d, one obtains a diagnostic function (Figure 7) that delineates the size regions where separation is maximal under a given set of experimental conditions.

The selection of experimental conditions should aim at placing a sample entirely within the "normal FFF" or the "steric FFF" regions to avoid the low resolution of the transition zone. In analyzing highly polydisperse samples, one must work at two or more field strengths when it is necessary to correct for mixed retention mechanisms (30). Samples of low polydispersity are readily accommodated in one region or the other, where their average sizes are determined from retention. In the general discussion of FFF zone broadening, it was noted that sample polydispersity adds a contribution (Equation 18) that is independent of carrier velocity (17). For

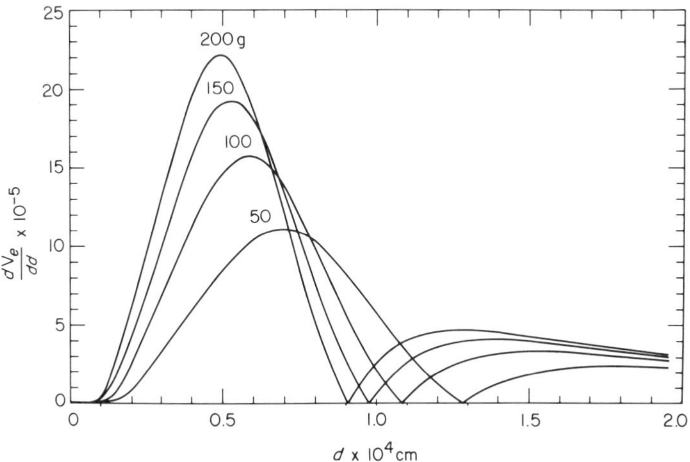

Figure 7. Variation in elution volume V_e with particle diameter. Log dV_e/dd is plotted as function of d to indicate regions of high and low selectivity at different field strengths (in gravities, g). Assumptions: $\Delta\rho$ = 0.05 g/cm^3, w = 0.0127 cm, and γ = 0.7.

well-retained samples, where $dR/d\lambda$ is constant by virtue of Equation 10, sedimentation FFF generates a polydispersity contribution to plate height that is a simple function of channel length L and the standard deviation σ_d in particle diameter within the sample (23):

$$\lim_{\lambda\to 0} H_p = 9L\left(\frac{\sigma_d}{d}\right)^2 \tag{28}$$

From plots of plate height versus carrier velocity for a selected set of experimental conditions, the composite of velocity-independent contributions (Equation 15) is the intercept at zero velocity (Figure 8). In work with well-characterized particle samples with known values for σ_d and d, it was possible to evaluate the magnitude of H_p under conditions where Equation 28 was applicable (21). In view of the generally good agreement between calculated H_p-values and intercepts from the plate height curves, it was concluded that nonideal zone-broadening effects, mainly caused by channel construction and operation, were insignificant.

Based on this result, we undertook to characterize a sample of bead-polymerized serum albumin suspended in ethanol (31). Figure 9 shows the sedimentation FFF fractogram recorded for this sample. From a knowledge of field strength and density difference between beads and ethanol, the average particle diameter was calculated, based on retention

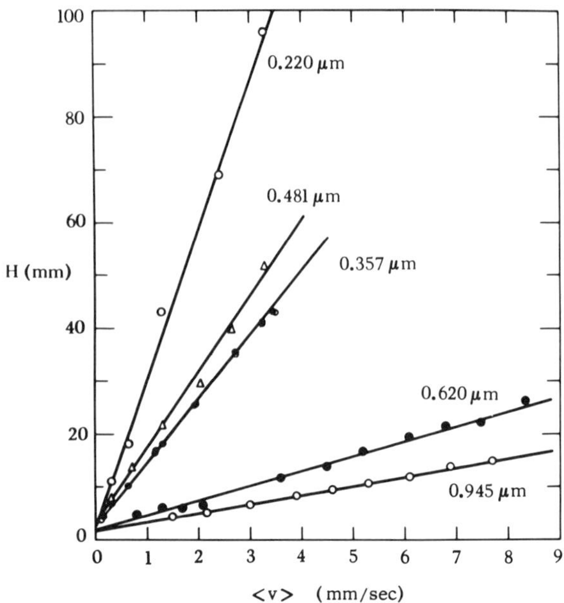

Figure 8. Plots of plate height H vs. mean flow velocity $\langle v \rangle$ for various polystyrene latex particles. The field strengths used for the different nominal particle sizes shown are: 168.7 g (0.220 μm), 21.5 g (0.481 μm), 66.7 g (0.357 μm), 31.0 g (0.620 μm), and 13.8 g (0.945 μm), where g is the gravitational acceleration in gravities. From reference 24.

at peak maximum. The plate height was determined graphically (32) from the peak width at half height, $w_{1/2}$, and the chart distance from start of flow to peak maximum, S:

$$H = \frac{L}{5.54} \cdot \left(\frac{w_{1/2}}{S} \right)^2 \qquad (29)$$

The carrier velocity during the run was used to evaluate the nonequilibrium contribution to plate height from Equations 20 and 21. The coefficient χ in Equation 20 was determined from the value of the parameter λ, which is related to the observed retention R at the peak maximum. With nonideal plate height terms assumed to be negligible, the polydispersity term H_p was directly calculated as the difference between the observed H and the nonequilibrium term H_n. Through Equation 27, H_p was converted into the corresponding standard deviation in particle diameter. Thus, in just over 1 hr, the sedimentation FFF characterization gave an average particle diameter of 0.349 μm with a standard deviation

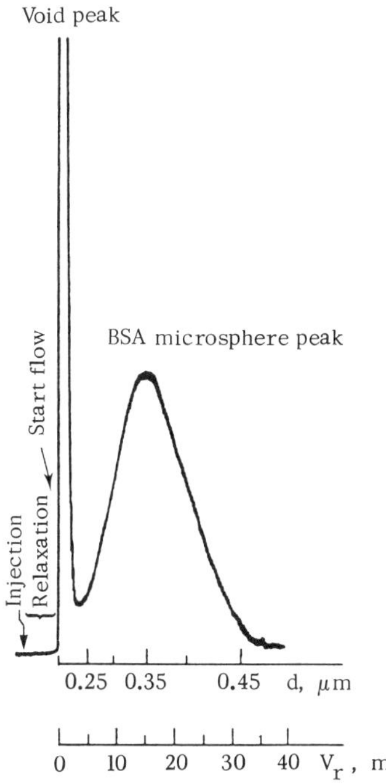

Figure 9. Sedimentation FFF fractogram of bovine serum albumin (BSA) microspheres. Field strength was 13.8 g and flow rate was 30 ml/hr. Retention volume and particle size scale (Equations 8, 9, and 24) are both shown on the horizontal axis. Reprinted from reference 31 with permission of the copyright owner.

of 0.040 μm. A more elaborate and much more time-consuming sizing by transmission electron microscopy showed an average diameter of 0.354 μm with a standard deviation of 0.062 μm.

Equation 29 gives an exact determination of plate height for Gaussian peaks. Since most real peaks (including the one in Figure 9) are non-Gaussian, plate heights are best calculated by moment analysis (33). The current computer interfacing (see below) of our sedimentation FFF equipment permits rapid evaluation of the first and second moments of a peak; this in turn gives more accurate determinations of average retention and plate height than the graphic technique used in reference 31.

The standard deviation in particle diameter is an inadequate characteristic of highly polydisperse samples, which are better described by their complete particle size distribution. The elution profiles recorded for such samples are themselves somewhat distorted representations of the mass or size distributions within each sample (Figure 10). As a first step in converting the fractogram into a distribution curve, elution volumes are reexpressed in terms of particle diameter, through the use of Equations 8 and 12. Second, one must correct the profile for the differences in selectivity that exist over the range of particle sizes present in the sample,

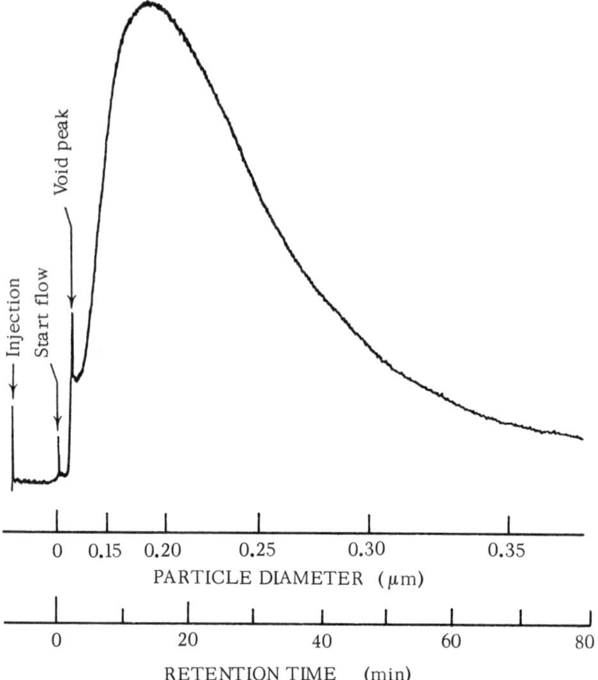

Figure 10. Typical sedimentation FFF fractogram of a highly polydisperse sample of PVC latex spheres. A particle size scale (Equation 8, 9, and 24) is superimposed on the elution time scale given by the chart recorder. Experimental conditions: field = 86.2 g, flow = 22.6 ml/hr, $\Delta\rho$ = 0.40 g/cm³, V^0 = 2.0 ml. From reference 30.

as illustrated in Figure 7. In regions of low selectivity, particles of a relatively wide range of diameters crowd into small volume elements and are together responsible for the detector signal. By contrast, regions of high selectivity yield relatively low particle contents in the effluent, and the detector response accounts more accurately for the concentration of particles in a narrow diameter range. In multiplying the detector response by the "scale correction factor," dV_e/dd (30), one corrects the fractogram for nonuniformities in selectivity.

A third transformation of the curve becomes necessary when the detector response is a function of both concentration and particle diameter. Optical detection is such a case, because the response to particulate samples stems mainly from light scattering (see Chapter 8). By application of either a theoretical (13, 30, 34) or an empirical scattering correction, the detector response can be made directly proportional to the concentration

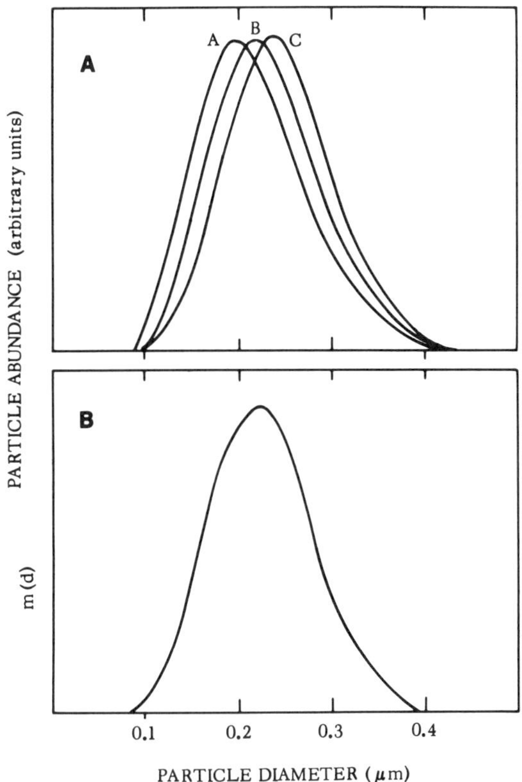

Figure 11. Size distribution curves for PVC latex by sedimentation FFF. Field = 193.9 g, flow = 28.5 ml/hr, V^0 = 2.0 ml. (A) A = with scattering correction only; B = without any correction; C = with scale correction only. (B) Fully corrected (with scale and scattering correction). From reference 30.

of particles in the effluent. Figure 11 demonstrates the effects of the two corrections. Here, the scattering correction, which leads to curve A in the figure, was performed using a computer routine developed by Barber and coworkers at the Department of Bioengineering, University of Utah, to account for Mie scattering (30). For many applications, it is not necessary to determine the exact size distribution for a sample. One can instead employ an uncorrected fractogram. Such is the case, for example, when the size distribution is used as a diagnostic of emulsion stability (35, 36). By following the uncorrected mass distribution (under a fixed set of experimental conditions) as it evolves with aging of the sample, one can readily determine any shifts in the composition of the emulsion.

2.2.2. Flow FFF

In FFF channels with semipermeable walls, a cross flow perpendicular to the regular channel flow may replace more conventional fields in causing fractionation. Under such circumstances, the "field induced velocity" U in Equation 2 is simply the rate of cross flow (expressed as linear velocity). Since U is the common velocity for all sample particles, differential retention is due solely to differences in the particle diffusion coefficient D. Equation 21 expresses D in terms of particle diameter d and reduced layer thickness λ. It thus has an inverse dependence on d (37):

$$\lambda_f = \frac{D}{Uw} = \frac{kT}{d \cdot 3\pi\eta \cdot U \cdot w} \tag{30}$$

For a channel with cross-sectional area A' (perpendicular to the cross flow), the velocity U is determined as the volumetric flow across the channel, $\dot{V}_c$, per unit area. Since volume V^0 equals the product of A' and channel thickness w, we may express λ_f in terms of the particle properties, D or d, and a set of easily measured experimental quantities (38):

$$\lambda_f = \frac{kT \cdot A'}{d \cdot \dot{V}_c \, w \cdot 3\pi\eta} = \frac{kTV^0}{d\dot{V}_c \, w^2 \cdot 3\pi\eta} = \frac{D \cdot V^0}{\dot{V}_c w^2} \tag{31}$$

As in the case of sedimentation FFF (Figure 3), the λ of flow FFF varies linearly with the inverse of field strength for a given sample (38).

Measured retentions are converted directly into sample diffusion coefficients through use of Equations 12 and 31. Figure 12 compiles a set of diffusion coefficients obtained from retention measurements on proteins and latex particles in a flow FFF unit. For the proteins, experimental D-values are plotted against diffusivities obtained from outside sources. The particle D-values were calculated using Equation 21 in conjunction with the manufacturer's values for particle diameter. The data cluster around a line with the ideal 45° slope (38).

Figure 13 shows the fractionation of a mixture of silica particles by flow FFF (39). The abscissa is graduated in effluent volume, and the elution position of one-column volume, V^0, is specifically indicated. The figure also contains a particle diameter scale calculated from Equations 9 and 31. The smaller particles elute at positions that are in reasonable agreement with the known average size of respective fraction, as indicated in the figure. The largest particle type within the sample, fraction "K-7" gives an abnormally broad peak, whose center falls at a larger average

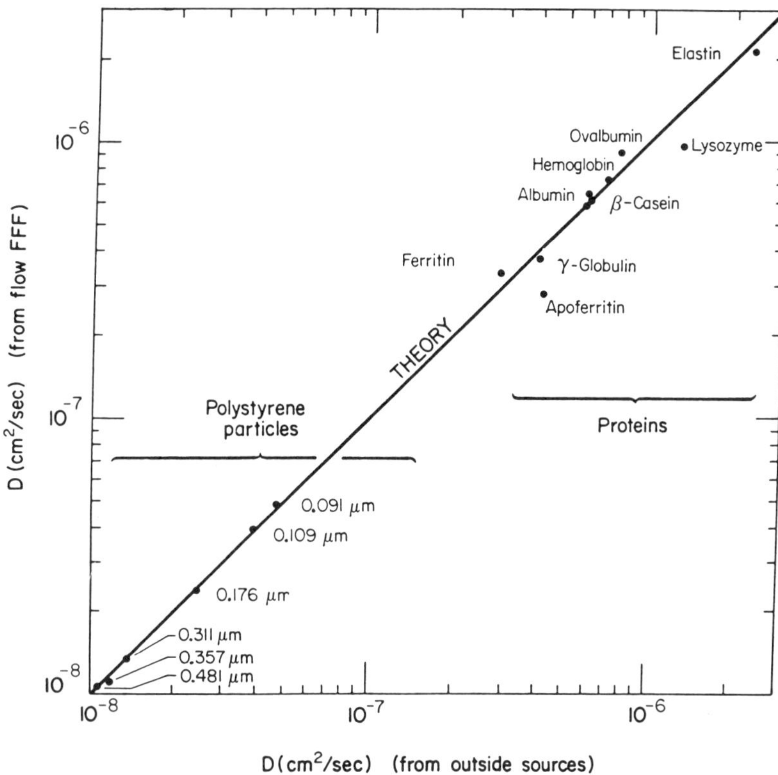

Figure 12.   A plot of D-values derived from flow FFF retention measurements (via Equation 31) vs. D-values acquired from independent sources. Reprinted from reference 38 with permission. Copyright 1976 American Chemical Society.

diameter than is calculated for this fraction from electron microscopy. No doubt this peak would change in shape if the detector response were properly corrected for light scattering and selectivity differences, as described above.

The early flow FFF channels showed excessive zone broadening, and deviations from theoretical predictions were seen in both slopes and intercepts of the plate height curves (38). A recent study of the zone broadening of proteins under flow FFF (40) indicated a marked improvement in performance. Diffusion coefficients derived from retention agreed closely with those calculated from the slopes of H versus $\langle v \rangle$ curves for seven samples, using Equations 19 and 20. Since the proteins were highly purified, and thus virtually monodisperse, any intercept would be due to nonideal effects, listed as ΣH_i in Equation 15. The observed values ranged

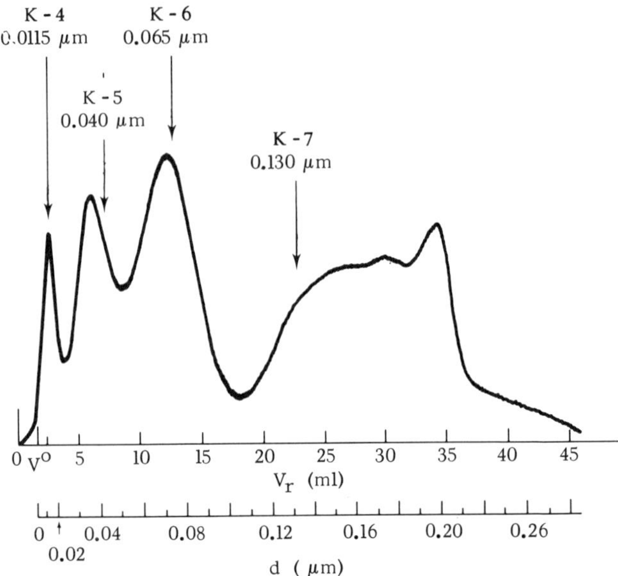

Figure 13. Flow FFF fractogram of silica particles. The sample was a mixture of fractions K-4, K-5, K-6, and K-7 obrained from J. J. Kirkland at DuPont de Nemours Company. The elution volume scale (V_r) is compared with the Stokes diameter scale obtained via Equation 31. Conditions were: field ($\dot{V}_c$) = 11.1 ml/hr, flow ($\langle v \rangle$) = 3.16 ml/hr, and V^0 = 1.85 ml. Reprinted from reference 39 with permission.

between 0.3 and 0.6 mm, which is 5–10% of the plate heights recorded at practical working flow rates (~20 ml/hr). These small values of the nonideal plate height contributions are decidedly satisfactory. When analyzing the zone broadening of moderately polydisperse samples in flow FFF, one can thus assume that H_p is the only important velocity-independent term in Equation 15. In this case, the intercept of the plate height curve arises from sample polydispersity alone.

In flow FFF, the term $d\lambda/d\bar{d}$ of Equation 18 equals $-(\lambda/d)$. Highly retained samples, for which $dR/d\lambda$ is constant, generate the following polydispersity contribution:

$$\lim_{\lambda \to 0} H_p = L\left(\frac{\sigma_d}{d}\right)^2 \tag{32}$$

The standard deviation in diameter, σ_d, depends on the square root of H_p in Equations 28 and 32. Thus, small errors made in treating the in-

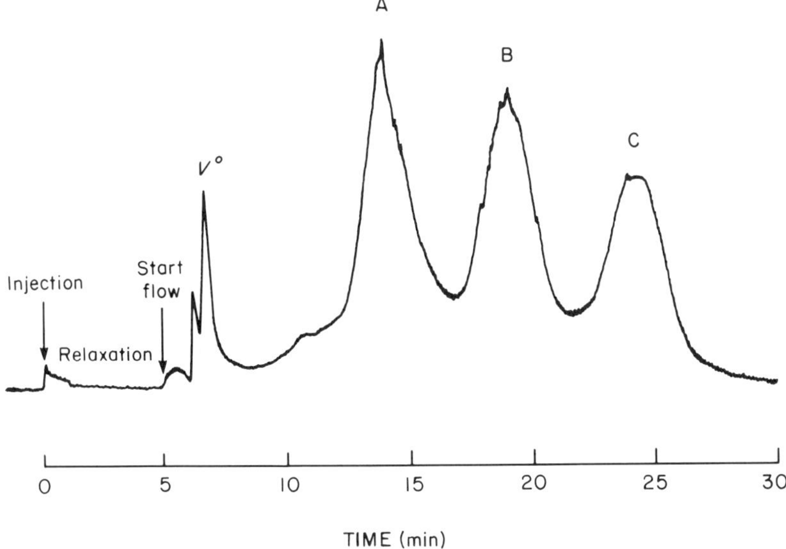

Figure 14. Steric FFF fractogram of silica particles. Peak A represents particles in the 10–14 μm size range; peak B depicts those of 7.5-μm average diameter; peak C illustrates those of 5.6-μm diameter. Separation took place in a horizontal channel under 1 g at a flow rate of 39.1 ml/hr. Reprinted from reference 41, by courtesy of Marcel Dekker, Inc.

tercept as caused entirely by sample polydispersity are reduced further by the power of $\frac{1}{2}$ in calculating σ_d.

2.2.3. Steric FFF

When particle diameters reach the layer thickness l (Equation 3) that one calculates for a zone of point masses, the zone as a whole will sample faster flow lines than its ideal counterpart. As a result, it will move downstream at a higher velocity. This condition is accounted for by Equation 12. For particles much larger than l, the zone no longer travels as a diffuse cloud, but each particle touches the accumulation wall. The zone velocity then depends on the degree to which the particles expand into regions of faster flow.

The elution order becomes reversed with respect to normal FFF, since large particles migrate downstream more rapidly than small ones (Figure 14). According to Equation 13, the retention factor R equals $6\gamma\alpha$, where α is the ratio of the particle radius to the channel thickness w. The factor γ accounts for a number of velocity-dependent effects, one of which is

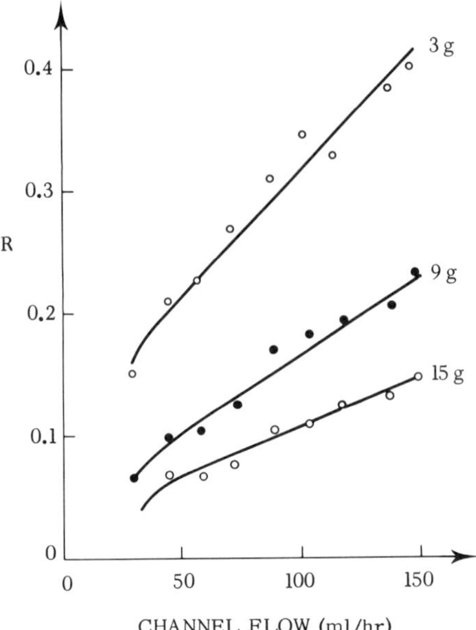

Figure 15. Steric FFF of human red blood cells. Variations in retention R with flow rate under different field strengths. From reference 44.

the frictional drag on the particle. As long as the particle is held in contact with the wall by the field, this frictional reduction in particle velocity gives γ a value of less than 1 (41). Because of the spatial extension of the particle through liquid layers of different velocities, the particle experiences a velocity-dependent lift force directed toward the faster flows in the center of the channel (42). As long as the field-induced settling force is larger than the lift force, the factor γ remains less than 1. The first steric FFF separations (17, 43) took place in flat, horizontal channels with gravity as the field. Dense particles, such as silica, were retained with γ-factors between 0.7 and 1; however, particles with densities close to that of the carrier [e.g. latex beads (42) or animal cells] eluted far ahead of their denser counterparts. By performing the fractionation in sedimentation FFF channels that were spun to a field 15 times gravity, the lift forces generated even by very high channel flows could be offset by the increased settling forces. Figure 15 illustrates how the retention of human red blood cells varies as a function of field and flow (44).

These cells are retained to the same degree under a flow of 150 ml/hr at 15 gravities as they are under 30 ml/hr at 3 gravities. This fact suggests that, if desired, an increased speed of fractionation at a fixed resolution can be acheived through an increase in the field that holds the particles in the wall region.

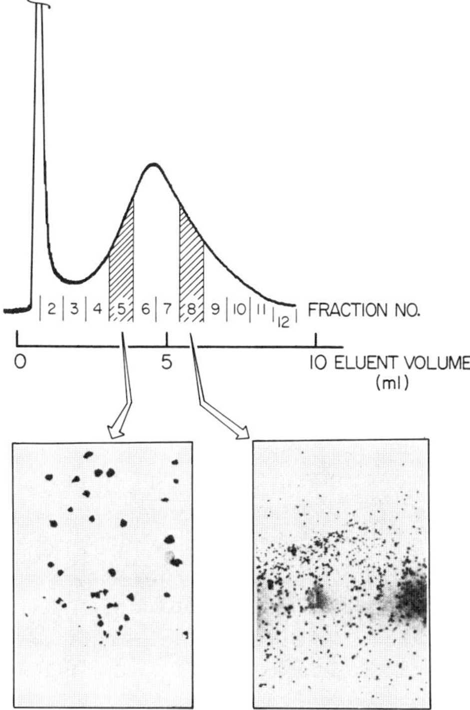

Figure 16. Characterization of residues from coal liquefaction. Steric FFF was performed in a horizontal channel under 1 g at a flow rate of 26 ml/hr. Fractions of 0.8 ml were collected. Microscopic examination of fractions 5 and 8 demonstrates the sizing. From reference 45.

Zone broadening in steric FFF occurs to a large extent through different mechanisms than in normal FFF, where the nonequilibrium effect plays a dominant role. When the steric effect is fully developed, all particles are thought to touch the wall at all times, thus eliminating mass transfer between liquid strata of different velocities. Indeed, experimentally observed plate heights vary only slowly with flow rate over a region of velocities where lift forces are compensated for by strong settling forces.

Much work is yet needed to fully understand the processes of retention and zone broadening in steric FFF. In the interim, the technique can be used for the separation and fingerprinting of particulate samples, although sizing at this point must involve the use of calibrated standards. Figure 16 demonstrates steric FFF characterization of a residue from coal liquefaction (45). To prove reproducibility of the separation, fractions 5 and 8 in the original run were reinjected into the system and emerged at positions corresponding to their original collection volumes. Microscopic analysis of these fractions clearly demonstrates that a sizing of the sample had occurred, since fraction 5 contained significantly larger particles than fraction 8. Steric FFF fractograms of coal liquefaction residues show substantial differences, depending on whether or not the coal was catalytically hydrogenated prior to the liquefaction process. Such results suggest the use of steric FFF for rapid identification of samples in process control work.

2.3. Programming Techniques

In view of the high resolution of the FFF techniques, and the broad range of sizes that may be present in a given sample, it is sometimes convenient to gradually change some experimental parameter during a run to optimize performance in time and resolution. The analysis usually starts at a high field strength capable of resolving the smallest components in the sample. If this high field were kept constant, it would give excessive retention to massive particles, and the analysis time would be inconveniently long. By gradually lowering the field, one may compress the fractogram in time without suffering an undue loss in resolution (26). A particularly attractive field programming has been introduced by Yau and Kirkland (13, 14) for sedimentation FFF. According to this protocol, named "Time Delayed Exponential Sedimentation FFF," the field is held constant for some time τ, after which it is led to decay exponentially with τ as time constant. This arrangement gives a straight proportionality between retention time t_r for a particle and the logarithm of its mass,

$$\ln m = \ln \mu + \frac{t_r}{\tau} \tag{33}$$

or diameter,

$$\ln d = \ln \beta + \frac{t_r}{3\tau} \tag{34}$$

where

$$\mu = \frac{6kT\tau}{e \cdot t^0 G_0 \cdot w \cdot \Delta\rho/\rho_s} \tag{35}$$

and

$$\beta = \frac{36kT\tau}{\pi e t^0 G_0 w \Delta\rho} \tag{36}$$

Here, e is the base of the natural logarithms, t^0 is the time required to sweep out one void volume, and G_0 is the initial field strength. All other quantities are as previously defined. The gain in analysis time offered by this approach, as compared to operation at constant field, is illustrated in Figure 17 for a mixture of polystyrene latex particles. Separations such as these have also been carried out in minutes rather than hours.

In the previous discussion of zone broadening, the nonequilibrium term was the only velocity-dependent factor of significance. High retention, with its heavy compression of particle zones (small λ values), has a dampening effect on nonequilibrium zone broadening by virtue of coefficient χ, which is closely approximated by $24\lambda^3$ for such zones (see Equations 16 and 17). These facts suggest a programmed increase in longitudinal flow at constant field as an alternative to field programming (46). Low velocities in the early phase of a separation help resolve poorly retained components, whereas more retained components can stand elution at progressively higher flow rates. Figure 18 illustrates this technique by a series of separations of polystyrene latex particles performed at constant field. The 4 ml/hr separation gave more than adequate resolution of the mixture, but required over 10 hr for completion. Under a flow of 32 ml/hr, it took 1.5 hr to elute the sample, but resolution of the less-retained components was poor. By allowing the flow rate to increase with the square of time from 4 ml/hr to 72 ml/hr, the sample was given an adequate resolution in just over an hour.

Thus, it is clear that a judicious choice of programming strategy is important in optimizing resolution and speed of analysis.

2.4. Sample Introduction

Field-flow fractionation is inherently an analytical technique whose accuracy is enhanced by low solute concentration and small injection volumes. To achieve optimum separation, it is particularly important to gauge the injected amounts to avoid nonideal particle interactions within the compressed zone. With current equipment design, we prefer to inject less than 1 mg of solids.

Injection volumes are ideally kept small ($<10\ \mu$l) to avoid undesired zone broadening, particularly in conjunction with polydispersity evaluation from plate height curves. Recently, however, we introduced a technique for handling large volumes of dilute sample (47). Rather than injecting, by means of a syringe, directly into the channel under flow but no field, the sample is now slowly pumped into the channel under a field that is considerably stronger than the "working field" at which the actual analysis takes place. Figure 19 demonstrates the principle of feed con-

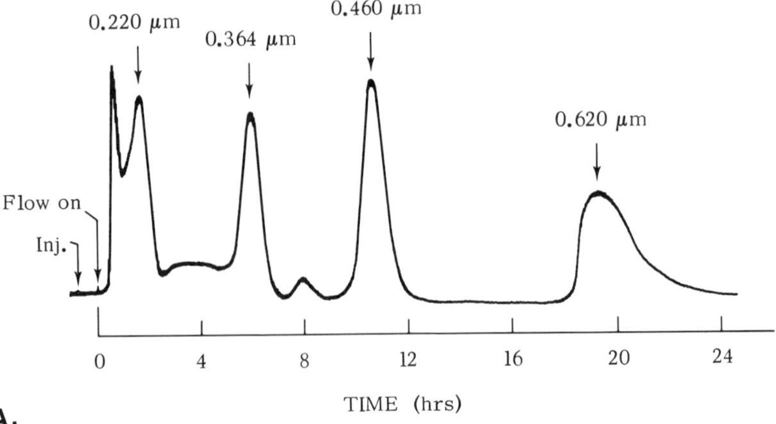

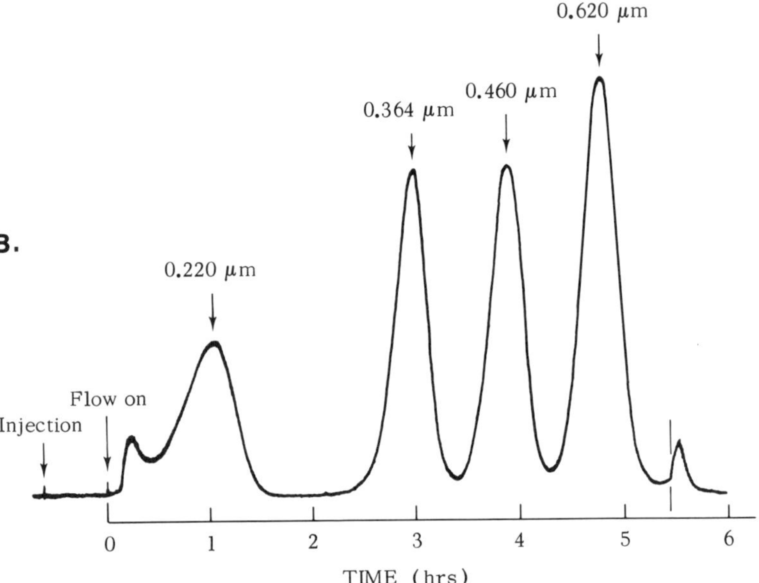

Figure 17. Time-delayed exponential programming in sedimentation FFF. (*A*) Separation of a mixture of polystyrene latex particles under a constant field of 66 *g*. The flow rate was 10.7 ml/hr. (*B*) Separation of the same mixture using a time-delayed exponential field decay. The initial field was 136 *g* and the time constant τ was 80 min. The flow rate was 30 ml/hr. From M. E. Hansen, University of Utah project, unpublished results.

238

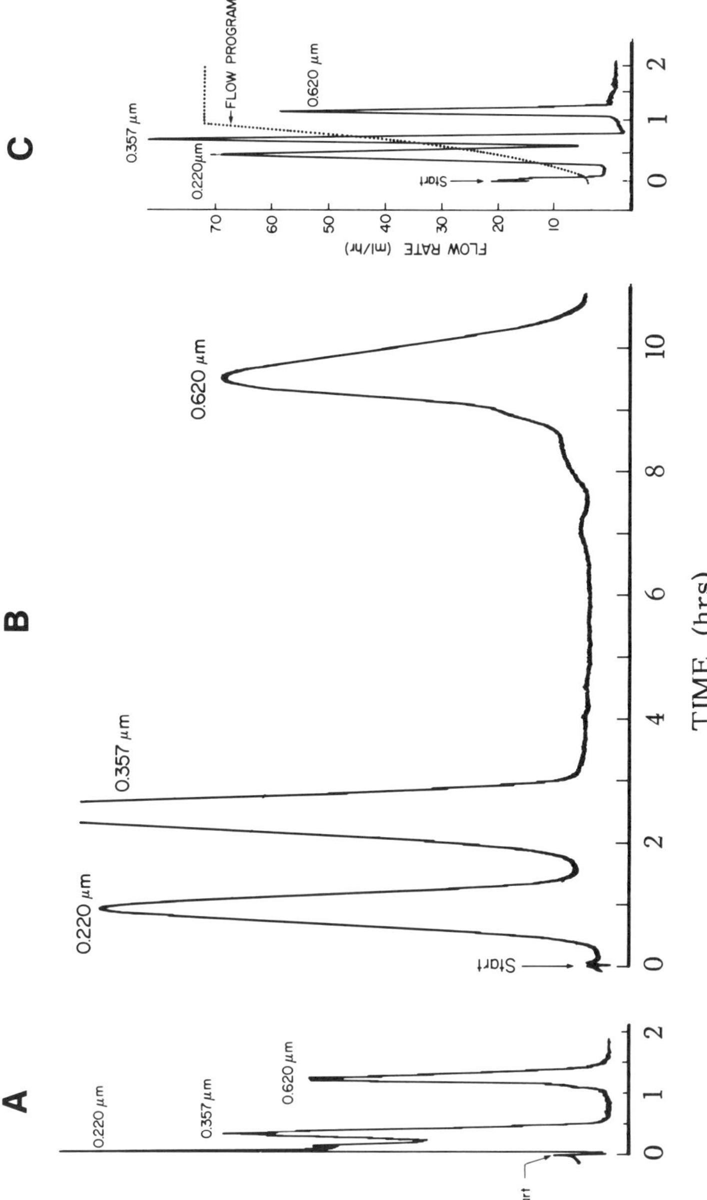

Figure 18. Effect of flow programming on the separation latex particles in sedimentation FFF. The field is held constant at 89.5 g. (*A*) Constant flow of 32 ml/hr. (*B*) Constant flow of 4 ml/hr. (*C*) Quadratic increase in flow rate with time. Start at 4 ml/hr and end at 72 ml/hr. Reprinted from reference 46 with permission. Copyright 1978 American Chemical Society.

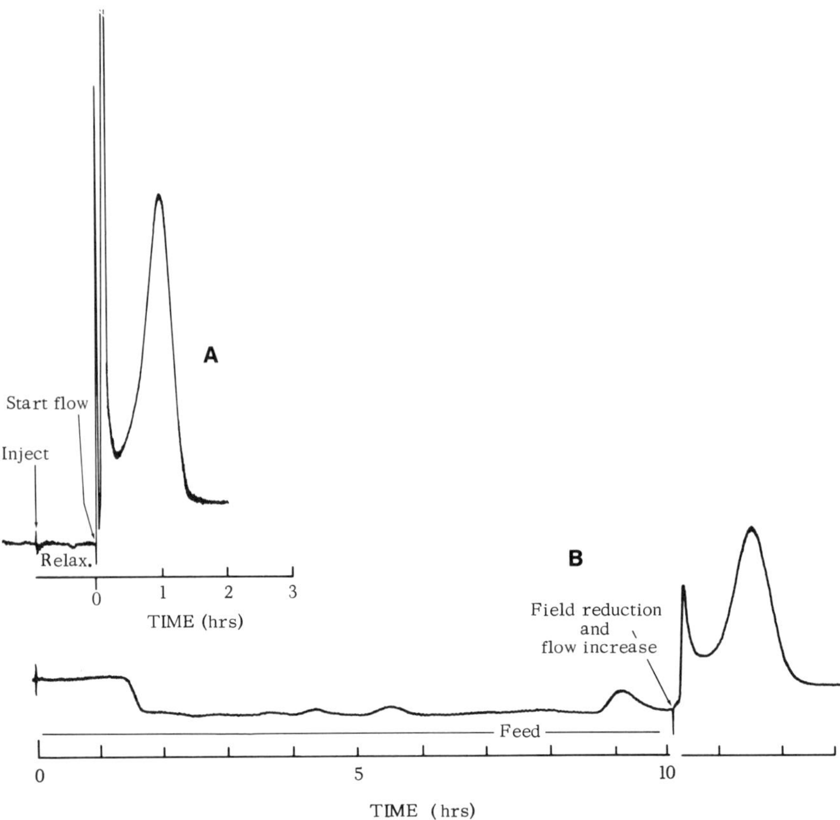

Figure 19.   Injection of large volumes of dilute sample. (*A*) Direct injection of a 1-μl sample of polystyrene latex particles with 0.357-μm diameter containing 0.1 mg solids. (*B*) Feed injection of the same amount of latex particles diluted to 50 ml. Conditions for the feed cycle were: field = 278.9 *g*, flow = 5.8 ml/hr. The separation was carried out at 66.7 *g* in both cases, with flows of 35.2 ml/hr (*A*) and 30.8 ml/hr (*B*). Reprinted from reference 47, by courtesy of Marcel Dekker, Inc.

centration, where fractograms from two runs at the same working field and flow are compared for the same amount of injected solids. In one case, a 1-μl sample was injected directly onto the head of the channel; in the other example, 50 ml of the sample (now in a 50,000-fold dilution) was pump-fed into the channel, which was spinning at 1800 rpm. Once the entire sample had entered the column, the field was reduced to its working level of 880 rpm and the flow was similarly adjusted to the working level of 30.8 ml/hr. Particles were retained to the same degree in both cases, and although the peak was noticeably broader for the feed-injected sample, it was still modest enough to prove that on-line concentration is a useful tool in the analysis of such dilute particle suspensions as stream waters (48). This sample injection technique was also described in reference 12 for the preparative separation of particles by sedimentation FFF.

3. INSTRUMENTATION

3.1. General Description

At the time of this writing, there are no commercially available sedimentation or flow type FFF instruments. However, high-performance sedimentation FFF was carried out by Kirkland, Yau, and coworkers at the DuPont Experimental Station (12, 13). Since prototype instruments for both subtechniques were developed in our laboratory, the following discussion centers on designs used at the University of Utah, with references to the significant advances made by the DuPont group.

Irrespective of subtechnique, the FFF channel or column is generally operated in line with a solvent pump, a detector, a chart recorder, and a fraction collector. In view of the open channel geometry, there is, for the most part, no need to use high-pressure pumps for delivery of the longitudinal flow, and peristaltic pumps are often adequate for this task. The detector is any unit designed for high-performance liquid chromatography. To keep nonideal zone broadening to a minimum, the detector volume should be 10 μl or less. In our laboratory, the signal from the detector is fed both to a strip chart recorder and to a PDP 11/03 minicomputer for processing at a later time. This laboratory computer is capable of simultaneous data acquisition from 16 units, with a minimum sampling interval of 0.2 sec. While sampling, the computer also controls the rotor speed of the sedimentation FFF units or the pumps that regulate cross flows in the flow FFF systems. At the end of a run, the collected data are transferred to a PDP 11/34 computer for evaluation of the first and second moments (i.e., the average retention and the plate height) of each peak identified in the fractogram. These data are in turn converted into the appropriate sample characteristics, as outlined above.

3.2. Specific Features

3.2.1. Sedimentation FFF

A detailed description of our sedimentation FFF instrument has been given previously (49). Briefly, the flow channel is cut from a sheet of steel

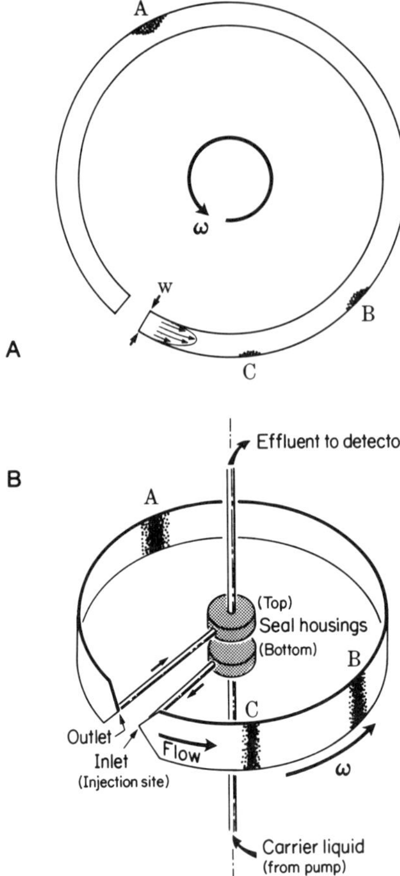

Figure 20.   Schematic representation of the sedimentation FFF channel with three zones A, B, and C displaying different migration velocities. (*A*) View along the axis of rotation. Channel thickness w is typically 0.01–0.02 cm. (*B*) View of channel from the side. The carrier liquid enters the spinning shaft at the bottom seal, is routed through the column, and on via the top seal to the stationary exit tube.

or Teflon whose thickness is uniform. The thickness of this spacer material determines the channel thickness w (see Figure 20). Flow channels are typically about 2 cm wide and just under a meter in length, and have both ends tapered at an angle of 60°. The spacer is sandwiched between two highly polished 0.63-cm ($\frac{1}{4}$ in.) thick strips of stainless steel, and the whole is either welded together (steel spacer) or held together mechanically (Teflon spacer). This latter unit is assembled by fitting the steel wall-spacer sandwich into a track in each of two solid aluminum rings, which gives the channel the proper curvature to fit inside an aluminum rotor basket. Once positioned inside the rings, the channel sides are sealed by tightening a set of 0.63-cm ($\frac{1}{4}$ in.) set screws, which perforate the inner walls of the rings and are positioned 1 in. apart. This construction makes possible easy assembly and dismantling of the unit (e.g., to change the surface composition of the channel walls).

The inlet and outlet are welded onto the inner channel wall, and the spacer is positioned to place these ports at the tips of its tapered regions. The inlet consists of a small steel cube that accommodates a septum injection port and the stainless steel tubing through which the carrier enters the channel just below the septum. The outlet opens directly into a 0.101 cm i.d. (0.40 in.) or smaller stainless steel tubing.

The carrier liquid enters the system through a stationary shaft, which seals against the rotating parts of the unit via a special self-lubricating "O" ring (formulation N-256 or 827 from Parker Seals, Lexington, KY) or a Teflon seal from Bal Seal Corp. (Tustin, CA) (31). The Teflon seal is needed for work with nonaqueous carriers. Once in the spinning part of the system, the liquid moves via an 0.051 cm (0.20 in.) i.d. tubing through the center shaft and out to the perimeter of the rotor basket, where the inlet to the channel is located. The effluent from the channel is likewise routed inward to the center shaft and through an "O" ring seal to a stationary exit shaft. These seals are of limited durability and permit spin rates of up to 3000 rpm. The spinning shaft carries a slotted disk and infrared emitter–sensor assembly for monitoring rates of rotation. The signal from this unit is fed to the laboratory computer, which controls the speed of the centrifuge rotor, and thus the field strength G in Equation 23.

Kirkland and coworkers (12, 13) use a face-seal, which enables them to reach considerably higher spin rates with their sedimentation unit. Recently (51), this group reported on the design of instrumentation capable of operating at 100,000 g. The column in this case consists of an inner, expandable plastic cylinder into which the channel is milled. Under the influence of the field, this inner cylinder seals directly against the polished wall of the metal rotor, thus creating a channel that is easily disassembled for inspection of its walls.

To reach the higher gravitational forces, channels will have to be spun in vacuum, as with commonly used ultracentrifugation equipment, for reduced friction between the rotor and its surroundings. The high operating speeds will also necessitate controlled cooling of the equipment.

Operation at increasingly higher fields will extend the sedimentation FFF separation range in the direction of lower molecular weights. While creating resolution of smaller masses, one simultaneously concentrates larger masses into extremely thin layers in the channel, where concentration anomalies may occur, and where particle–wall interactions may become significant. It will thus be important to pay close attention to the

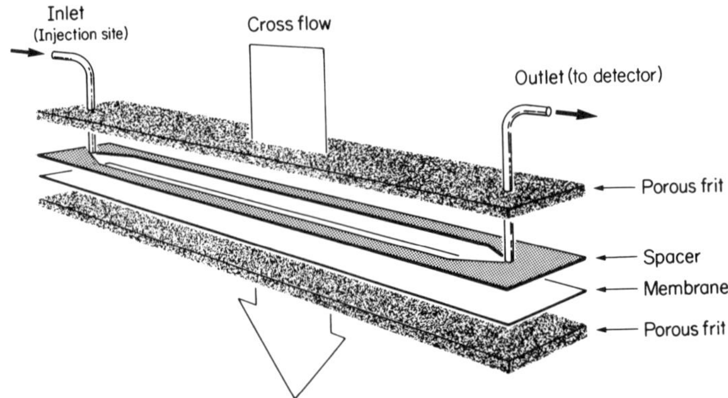

Figure 21. Schematic representation of the flow FFF apparatus. The spacer defines the channel geometry, including thickness w. Carrier liquid is forced to flow from an upper reservoir (not shown) across the upper frit and into the channel. Because of the exact match of the "inlet" and "outlet" (longitudinal) flows, liquid is then forced to exit through the semipermeable membrane and its supporting porous frit to a lower reservoir (not shown).

chemical compositions of both carrier medium and channel surface, to minimize such interactions in work with highly retained samples.

3.2.2. Flow FFF

Reference 49 gives a detailed description of the flow FFF system. The channel geometry is similar to the one used for sedimentation FFF, but channels are generally only about 50 cm in length (see Figure 21). The Teflon spacer is clamped between two plexiglass or steel blocks, which are hollow and are equipped with a porous frit on one of the sides. The frits are polyethylene or steel, respectively, and face the interior of the channel in the assembly. Between the bottom frit and the spacer is a membrane, such as Millipore's PTGC skin membrane (Millipore, Bedford, MA), with a molecular weight cut-off limit of around 10,000 dalton. This membrane has been suitable for work in both aqueous and certain non-aqueous solvents. Against this membrane wall, the accumulation of sample particles is forced to take place. The opposite channel wall needs no membrane, but the frit must have such resistance to flow that the pressure drop across this wall is significantly larger than the pressure drop along the length of the channel. If this condition is not fulfilled, the content of the channel is forced up into the frit in the entrance region, whereas the cross flow overpowers this effect closer to the exit.

The relative magnitudes of the longitudinal and cross flows are maintained by a controlled delivery and withdrawal of liquid at the entrance and exit of the channel. Under modest channel pressures, this is conveniently handled by a bifunctional peristaltic pump. At elevated pressures, one is forced to operate a restriction valve at the end of the channel to identically match the two flows. With the longitudinal flow under control, the field strength $\dot{V}_c$ in Equation 31 is identical to the pump rate of the cross-flow pump. This pump, in turn, can be computer controlled, should field programming be desired.

3.2.3. Steric FFF

Since steric FFF constitutes a limiting case of normal FFF that becomes operational for large particles, it follows that separations of this type take place in the "normal" equipment mentioned earlier. For large and dense particles, earth's gravity is large enough to offset the velocity-induced lift forces described above. In such cases, a horizontal channel, with a Teflon spacer clamped between two glass plates (49), is a simple but adequate separation tool. The glass plates provide the desired flatness of the walls and, in some cases, permit visual inspection of the run. Owing to the ease with which this system is assembled, one can readily modify the channel surfaces, by polymer coating or chemical derivatization of the glass, for example. In this manner, surface adsorption may be reduced or enhanced for maximum selectivity and recovery.

4. CONCLUSION

The FFF technique in either its sedimentation or flow modification, is capable of characterizing particles over an unusually broad size range. With about 1 nm (corresponding to a molecular weight of 1000) as the lower limit of resolution by flow FFF, 10 nm (or 500,000 molecular weight) for sedimentation FFF, and an upper limit for both techniques (in the steric regime) of over 100 μm, this resolution range spans 5 orders of magnitude in sample diameter.

Within the wide resolution range, there are regions where a complex sample is more or less efficiently resolved. Of special concern is the transition region between normal and steric operation, where the elution ceases to increase with particle size and instead begins to decrease as larger particles are fractionated. In analyzing samples of very broad size distribution, one must consequently carry out the fractionations at a series of different field strengths, so that the lower resolution "transition" region is moved up or down the particle size spectrum (see Figure 7). Only with such systematic variations in field strength can one utilize the full FFF resolution range.

Under "normal" operation, the sedimentation FFF offers higher size selectivity than its flow counterpart and should be the technique of choice, provided sample particles are large enough to be retained by the available equipment. Although its resolution is less favorable, the flow FFF has an advantage over the sedimentation analogue in the interpretation of elution data, since its retention is a function of only one particle property (diffusivity), rather than two for sedimentation FFF (molecular weight or diameter, and density).

Of particular advantage for particle sizing is the wide latitude offered by FFF in the choice of carrier medium for the analysis. Unlike the Coulter counter, which requires an ionic suspension medium for sizing, and density gradient sedimentations, which require additions of density modifiers in large concentrations, the carrier in FFF is selected for maximum sample compatibility. In analyzing particle suspensions, one can thus work directly with the suspension medium as carrier. In this context, FFF has a clear advantage over electron microscopic techniques, which require drying and mounting of the sample prior to examination.

Unlike other particle characterization techniques, such as microscopy, light scattering, and Coulter counting, sizing by FFF is an elution procedure, where sample components emerge at the channel exit in the order specified by their physical properties. The sorted particles are then easily collected and subjected to further identification (e.g., with respect to biological activity, morphology, or chemical composition), thus adding another dimension to the analytical information obtainable for polydisperse samples.

ACKNOWLEDGMENT

The support by NIH grant GM 10851-24 is gratefully acknowledged.

NOTATION

A Constant in the plate height equation.

A' Cross-sectional area of channel.

c Particle concentration.

$c(x)$ Concentration of particles at a distance x from the wall of the channel.

C Constant in the plate height equation.

d Particle diameter.

$\bar{d}$ Average particle diameter.

D Diffusion coefficient of particles.

f Frictional coefficient.

F Force exerted by applied field.

g Gravitational field, G/gravitational force (980.66 cm/sec^2).

G Gravitational acceleration, $\omega^2 r'$.

G_0 Initial field strength used in time-delayed exponential sedimentation FFF.

H Total plate height.

H_l Plate height term caused by axial diffusion.

H_n Nonequilibrium plate height term resulting from resistance to mass transfer between velocity streamlines.

H_p Plate height term resulting from polydispersity of sample.

ΣH_i Composite plate height term consisting of extracolumn effects.

k Boltzmann constant.

l Layer thickness parameter, D/U or kT/F.

L Channel length.

m Mass of a particle.

m' Effective mass of a particle adjusted for the buoyancy of the medium.

r Particle radius.

r' Radius of rotor.

R Retention factor, $v_{\text{zone}}/\langle v \rangle$.

S Distance to peak maximum.

t^0 Elution time for unretained component.

t_e Elution time for retained component.

T Temperature, Kelvin.

U Field induced velocity prependicular to flow.

v_{zone} Zone velocity.

$v(x)$ Velocity profile.

$\langle v \rangle$ Average linear velocity.

V^0 Elution volume for unretained component (column or channel volume).

V_e Elution volume for retained component.

$\dot{V}_c$ Volumetric flow across channel.

w — Channel thickness.

$w_{1/2}$ — Peak width at half height.

x — Distance from the wall of the channel.

α — Parameter equal to r/w.

γ — Velocity-dependent factor, close to unity.

η — Viscosity of suspension medium.

λ — Reduced layer thickness, l/w.

λ_f — Reduced layer thickness in flow FFF.

λ_s — Reduced layer thickness in sedimentation FFF.

ρ — Density of solvent.

ρ_p — Density of particle.

$\Delta\rho$ — Density difference between particle and suspension medium.

σ_d — Standard deviation of the particle diameter.

σ^2 — Variance of sample zone.

τ — Time constant used in time-delayed exponential sedimentation FFF.

χ — Nonequilibrium, plate height coefficient.

ω — Angular velocity of a rotor.

REFERENCES

1. G. H. Thompson, M. N. Myers, and J. C. Giddings, *Sep. Sci.*, **2**, 797 (1967).
2. J. C. Giddings, *J. Chem. Phys.*, **49**, 81 (1968).
3. K. D. Caldwell, L. F. Kesner, M. N. Myers, and J. C. Giddings, *Science*, **176**, 296 (1972).
4. J. C. Giddings, M. Martin, and M. N. Myers, *J. Chromatogr.*, **149**, 501 (1978).
5. J. C. Giddings, F. J. Yang, and M. N. Myers, *Sep. Sci.*, **12**, 331 (1977).
6. J. C. Giddings, F. J. F. Yang, and M. N. Myers, *Sep. Sci.*, **10**, 133 (1975).
7. J. C. Giddings, F. J. Yang, and M. N. Myers, *Science*, **193**, 1244 (1976).
8. E. N. Lightfoot, P. T. Noble, A. S. Chiang, and T. A. Ugulini, *Sep. Sci. Technol.*, **16**, 619 (1981).
9. J. Janca and K. Kleparnik, *Sep. Sci. Technol.*, **16**, 657 (1981).
10.. M. Martin and R. Reynaud, *Anal. Chem.*, **52**, 2293 (1980).
11. J. J. Kirkland, W. W. Yau, and F. C. Szoka, *Science*, **215**, 296 (1982).
12. J. J. Kirkland, W. W. Yau, W. A. Doerner, and J. W. Grant, *Anal. Chem.*, **52**, 1944 (1980).

13. J. J. Kirkland, S. W. Rementer, and W. W. Yau, *Anal. Chem.*, **53**, 1730 (1981).

14. W. W. Yau and J. J. Kirkland, *Sep. Sci. Technol.*, **16**, 577 (1981).

15. J. C. Giddings, S. R. Fisher, and M. N. Myers, *Am. Lab.*, **10**(5), 15 (1978).

16. M. E. Hovingh, G. H. Thompson, and J. C. Giddings, *Anal. Chem.*, **42**, 195 (1970).

17. J. C. Giddings and M. N. Myers, *Sep. Sci. Technol.*, **13**, 637 (1978).

18. J. Happel and H. Brenner, *Low Reynolds Number Hydrodynamics*, Prentice Hall, Englewood Cliffs, N.J., 1965, p. 34.

19. J. C. Giddings, *Dynamics in Chromatography*, Marcel Dekker, New York, 1965, Chapter 1.

20. J. C. Giddings, *Sep. Sci. Technol.*, **13**, 241 (1978).

21. G. Karaiskakis, M. N. Myers, K. D. Caldwell, and J. C. Giddings, *Anal. Chem.*, **53**, 1314 (1981).

22. J. C. Giddings, Y. H. Yoon, K. D. Caldwell, M. N. Myers, and M. E. Hovingh, *Sep. Sci.*, **10**, 447 (1975).

23. J. C. Giddings, F. J. F. Yang, and M. N. Myers, *Anal. Chem.*, **46**, 1917 (1974).

24. J. C. Giddings, G. Karaiskakis, K. D. Caldwell, and M. N. Myers, *J. Colloid Interface Sci.*, **92**, 66 (1983).

25. T. Svedberg and K. Pedersen, *The Ultracentrifuge*, Clarendon Press, Oxford, 1940.

26. F. J. F. Yang, M. N. Myers, and J. C. Giddings, *Anal. Chem.*, **46**, 1924 (1974).

27. J. C. Giddings, G. Karaiskakis, and K. D. Caldwell, *Sep. Sci. Technol.*, **16**, 607 (1981).

28. K. D. Caldwell, G. Karaiskakis, and J. C. Giddings, *J. Chromatogr.*, **215**, 323 (1981).

29. J. C. Giddings and M. N. Myers, *Anal. Chem.*, **54**, 2284 (1982).

30. F. S. Yang, K. D. Caldwell, and J. C. Giddings, *J. Colloid Interface Sci.*, **92**, 81 (1983).

31. K. D. Caldwell, G. Karaiskakis, M. N. Myers, and J. C. Giddings, *J. Pharm. Sci.*, **70**, 1350 (1981).

32. L. R. Snyder and J. J. Kirkland, *Introduction to Modern Liquid Chromatography*, 2nd ed., Wiley-Interscience, New York, 1979, Chapter 2.

33. E. Grushka, M. N. Myers, P. D. Shettler, and J. C. Giddings, *Anal. Chem.*, **41**, 889 (1969).

34. C. A. Silebi and A. J. McHugh, in P. Becher and M. N. Yudenfreund (Eds.), *Emulsions, Latices, and Dispersions*, Marcel Dekker, New York, 1978, p. 155.

35. F. S. Yang, K. D. Caldwell, M. N. Myers, and J. C. Giddings, *J. Colloid Interface Sci.*, **93**, 115 (1983).

250   FIELD-FLOW FRACTIONATION OF PARTICLES

36. F. S. Yang, K. D. Caldwell, J. C. Giddings, and L. Astle, *Anal. Biochem*, in press.
37. J. C. Giddings, M. N. Myers, F. J. F. Yang, and L. K. Smith, in M. Kerker (Ed.), *Colloid and Interface Science*, Volume IV, Academic Press, New York, 1976, p. 381.
38. J. C. Giddings, F. J. Yang, and M. N. Myers, *Anal. Chem.*, **48,** 1126 (1976).
39. J. C. Giddings, G. C. Lin, and M. N. Myers, *J. Colloid Interface Sci.*, **65,** 67 (1978).
40. G. Karaiskakis, K. D. Caldwell and J. C. Giddings, manuscript in preparation.
41. K. D. Caldwell, T. T. Nguyen, M. N. Myers, and J. C. Giddings, *Sep. Sci. Technol.*, **14,** 935 (1979).
42. P. G. Saffman, *J. Fluid Mech.*, **22,** 385 (1965).
43. J. C. Giddings, M. N. Myers, K. D. Caldwell, and J. W. Pav, *J. Chromatogr.*, **185,** 261 (1979).
44. K. D. Caldwell, Z-Q. Cheng, P. Hradecky and J. C. Giddings, *Cells Biophysics*, submitted for publication.
45. H. Meng, K. D. Caldwell, and J. C. Giddings, *Fuel Processing Technology*, in press.
46. J. C. Giddings, K. D. Caldwell, J. F. Moellmer, T. H. Dickinson, M. N. Myers, and M. Martin, *Anal. Chem.*, **51,** 30 (1979).
47. J. C. Giddings, G. Karaiskakis, and K. D. Caldwell, *Sep. Sci. Technol.*, **16,** 725 (1981).
48. G. Karaiskakis, K. A. Graff, K. D. Caldwell, and J. C. Giddings, *Int. J. Environ. Anal. Chem.*, **12,** 1 (1982).
49. J. C. Giddings, M. N. Myers, K. D. Caldwell, and S. R. Fisher, in D. Glick (Ed.), *Methods of Biochemical Analysis*, Vol. 26, John Wiley, New York, 1980, p. 79.
50. J. C. Giddings, *J. Chem. Ed.*, **50,** 667 (1973).
51. J. J. Kirkland, C. H. Dilks, Jr., and W. W. Yau, *J. Chromatogr.*, **255,** 255 (1983).

DETECTION SYSTEMS FOR PARTICLE CHROMATOGRAPHY

ARCHIE HAMIELEC

Department of Chemical Engineering
McMaster University
Hamilton, Ontario, Canada

1. INTRODUCTION

Particle chromatography using packed beds has attracted the attention of those interested in measuring particle size distribution of spherical particles in the submicron range. It is particularly powerful for particles less than about 2500 Å in diameter, where losses owing to the holdup of particles in the packing is usually negligible (1).

One of the major problems with this form of chromatography is the large amount of axial dispersion (or peak broadening) that occurs with the larger particles in the sample. The contents of the detector cell are not monodispersed, but rather have a distribution that is likely to be unimodal, and is often quite broad. This distribution of particle sizes in the detector cell can greatly complicate the interpretation of the detector response and can lead to appreciable error in the particle size distribution and particle diameter averages, if not taken into account. In this regard, it is important to point out the advantages of different detector systems, such as those based on turbidimetric and differential refractometry detection. The promise of detection systems based on turbidity–spectra and quasi-elastic light scattering should also be discussed. These topics are treated in this chapter after a methodology of detector response interpretation is developed.

2. GENERAL RELATIONSHIPS

2.1. General Detector

The response for a general detector $F(v)$ is given by

$$F(v) = \int_0^\infty W(v, y)\, dy \tag{1}$$

where

$$W(v, y) \propto N(v, y)D^\gamma(y) \qquad \text{(Type 1)} \tag{1a}$$

or

$$W(v, y) \propto N(v, y)D^2(y)K(y) \qquad \text{(Type 2)} \tag{1b}$$

and where $W(v, y)\, dv\, dy$ is the area under the detector response in the retention volume range $v - (v + dv)$ caused by particles with mean retention volume in the range $y - (y + dy)$; $N(v, y)\, dv\, dy$ is the number of particles and $D(y)$ is the diameter of particles with mean retention volume y. Type 1 represents both the refractive index detector ($\gamma = 3$) and the turbidity detector in the Rayleigh scattering regime ($\gamma = 6$). Type 2 includes the turbidity detector in the Mie scattering regime, where $K(y)$ is the extinction coefficient for particles of diameter $D(y)$.

2.1.1. *Properties of the Detector Cell Contents*

Diameter averages in the detector cell are given by the following equations (2) in which D_N, D_W, D_S, D_{SS}, D_V, and D_τ are the number, weight, surface, specific surface, volume, and turbidity average particle diameter, respectively.

$$D_N(v) = \frac{\int_0^\infty D(y)N(v, y)\, dy}{\int_0^\infty N(v, y)\, dy} \tag{2}$$

$$D_W(v) = \frac{\int_0^\infty D^4(y)N(v, y)\, dy}{\int_0^\infty D^3(y)N(v, y)\, dy} \tag{3}$$

$$D_S(v) = \left[\frac{\int_0^\infty D^2(y)N(v, y)\,dy}{\int_0^\infty N(v, y)\,dy}\right]^{1/2} \tag{4}$$

$$D_{SS}(v) = \frac{\int_0^\infty D^3(y)N(v, y)\,dy}{\int_0^\infty D^2(y)N(y, y)\,dy} \tag{5}$$

$$D_V(v) = \left[\frac{\int_0^\infty D^3(y)N(v, y)\,dy}{\int_0^\infty N(v, y)\,dy}\right]^{1/3} \tag{6}$$

$$D_\tau(v) = \left[\frac{\int_0^\infty D^6(y)N(v, y)\,dy}{\int_0^\infty D^3(y)N(v, y)\,dy}\right]^{1/3} \tag{7}$$

The frequency distribution in the detector cell at retention volume v is given by

$$f(v, D)\,dD = \frac{-N(v, y)\,dy}{\int_0^\infty N(v, y)\,dy} \tag{8}$$

The negative sign accounts for the negative slope of the particle diameter–retention volume calibration curve.

2.1.2. Whole Sample Properties

The frequency distribution of the whole sample injected into the chromatograph is given by

$$f(D)\,dD = \frac{-\left\{\left[\int_0^\infty N(v, y)\,dv\right]dy\right\}}{\int_0^\infty \int_0^\infty N(v, y)\,dv\,dy} \tag{9}$$

The diameter averages for the whole sample can be found by integrating $f(D)$ in the usual manner.

2.1.3. Turbidimetric Detection

In the absence of multiple scattering, the turbidity of a spherical suspension is given by

$$\tau = \frac{1}{l} \ln\left(\frac{I_0}{I}\right) = N\frac{\pi}{4} D^2 K \tag{10}$$

where

l is the optical path length, cm.

I_0 is the intensity of the incident beam.

I is the intensity of the emerging beam.

N is the particle concentration, number/cm^3.

D is the particle diameter, cm.

K is the extinction coefficient, dimensionless. It is a function of the two dimensionless parameters $\alpha = \pi D/\lambda$ and m, where λ is the wavelength of light in the medium and m is the ratio of the refractive indices of particle and medium.

Turbidity is related to the absorbance, A, by

$$\tau = \frac{2.303}{l} \cdot A \tag{11}$$

The extinction coefficient may be calculated from light scattering theory after Mie (3, 4). In the Rayleigh scattering regime, the extinction coefficient for nonabsorbing particles is given analytically as

$$K = \frac{8}{3}\left(\frac{m^2 - 1}{m^2 + 2}\right)^2 \alpha^4 \tag{12}$$

The turbidity of very small particles (small relative to the wavelength of light) is proportional to the sixth power ($\gamma = 6$) of the particle diameter (Rayleigh scattering regime $\alpha < 1$): for larger particles, the exponent is smaller than six. As a consequence, the small particle signal is relatively weak, though it can be augmented by using shorter wavelengths. However, to obtain particle size distributions, the relative signal, rather than the absolute signal, is of interest. Calculations by Silebi and McHugh (5) indicate that a change of wavelength or refractive index ratio has a small effect on the relative signal for nonabsorbing particles. However, the relative signal is considerably improved at absorption wavelengths owing to a significant enhancement in the extinction coefficient of small particles. These theoretical deductions were experimentally substantiated by Nagy (6, 7), using polystyrene latices.

Appreciable error may result "if instruments which are suitable for ordinary absorption measurements are used for turbidity measurements without proper modifications and precautions" (8). Although such errors were believed to be negligible by earlier workers (5), their existence was unequivocally demonstrated by Husain et al. (9), who compared the detector response of a suspension of polystyrene spheres to that of a sodium dichromate solution. Furthermore, it was shown that impurities (such as residual styrene monomer in the polystyrene particles) and additives (such as emulsifier) can cause the measured extinction coefficients to differ from theoretical values based on Mie theory. The discrepancy may theoretically be accounted for by employing an effective imaginary refractive index ratio (6, 7).

2.1.4. Differential Refractometry Detection

Based on Mie theory, Zimm and Dandliker (10) derived a general refractive index expression given by

$$\frac{d\eta_s}{dc} = \frac{3\eta_m}{2\alpha^3\rho_p} \operatorname{Re}\left[\sum_{n=1}^{\infty} \frac{2n+1}{2n(n+1)} (a_n - b_n)\right] \tag{13}$$

where c is the weight concentration in g/cm^3; ρ_p is the particle density; α is a dimensionless size parameter ($\alpha = \pi D/\lambda$, where D and λ are, respectively, the particle diameter and wavelength in the medium); η_m and η_s are the refractive indices of the medium and dispersion, respectively; and a_n and b_n are functions of α and m (m is the refractive index ratio of particle to medium). The real value of the bracketed term is denoted Re. The equation given above does not contain the restriction that α be small. In the limit as $\alpha \to 0$, it reduces to

$$\frac{d\eta_s}{dc} = \frac{3\eta_m}{2\rho_p} \frac{(m^2 - 1)}{(m^2 + 2)} \tag{14}$$

In accordance with Equations 13 and 14, $d\eta_s/dc$ is expected to be independent of c and, at small values of α, to be independent of α as well.

Measurements by Silebi and McHugh (5) show a surprising agreement of measured data with Equation (14) for polystyrene latices as large as

350 nm. Subsequent data measured by Nagy (6) from the same laboratory indicate that $d\eta_s/dc$ reverses in sign with increasing particle size, implying that the signal is null for some intermediate particle size. Coll and Fague (11) observed that $d\eta_s/dc$ was independent of c for a given latex, though its value increased linearly with particle diameter. Neither Nagy nor Coll and Fague were able to explain their results satisfactorily. Interpretation of their data is complicated because of the use of a broad wavelength source.

It is the author's opinion, that these seemingly conflicting data are in fact consistent with the Zimm and Dandliker equation. Calculations indicate that, depending on the values of m and α, $d\eta_s/dc$ may either increase with particle size or decrease and eventually change sign (10, 12).

Differential refractometry shows a less dramatic dependence on particle size than turbidimetry of nonabsorbing particles. The advantage is, however, negated by the requirement of a higher sample concentration compared to the amount necessary for a photometric detection. This is due to the limited sensitivity of commercially available refractometers. The treatment of a refractometer detector response for axial dispersion, herein, is based on the assumption that $d\eta_s/dc$ is independent of particle diameter. In certain instances, as mentioned above, this assumption may not be valid.

3. INTEGRAL EQUATION FOR AXIAL DISPERSION

3.1. Introduction

All rigorous methods of correcting detector responses for peak broadening use the following integral equation as the basis

$$F(v) = \int_0^\infty W(y)G(v, y)\, dy \tag{15}$$

where $F(v)$ is the detector response at retention volume v, $G(v, y)$ is the normalized detector response for a particle of diameter $D(y)$, $G(v, y)$ is called the instrumental spreading function (or just the spreading function), $W(y)\, dy$ is the area under the detector response caused by particles of diameter $D(y)$, and $W(y)$ is called the detector response corrected for dispersion. Equation 15 is a Fredholme integral equation of the first kind and has been used extensively in various science and engineering applications.

Comparing Equations 1 and 15, it is clear that

$$W(v, y) = W(y)G(v, y) \tag{16}$$

$$W(y) = \int_0^\infty W(v, y)\, dv \tag{17}$$

Equations 16 and 17 will be used to derive correction equations for dispersion in the detector cell.

3.2. Limiting Forms of the Integral Equation

3.2.1. Uniform Spreading Function

The instrumental spreading function is uniform when its shape parameters are independent of retention volume; in other words, the shape parameters are the same for particles of different diameter. The integral equation now takes the form of a convolution integral,

$$F(v) = \int_0^\infty W(y)G(v - y)\, dy \tag{18}$$

This limiting form should be valid for samples with relatively narrow particle size distributions. Bilateral LaPlace transformation of Equation 18 gives (13, 14)

$$\overline{F}(v) = \overline{W}(s)\overline{G}(s) \tag{19}$$

This algebraic equation will be used to develop corrections for dispersion for the whole sample particle diameter averages.

For the case of a uniform Gaussian spreading function, Equation 18 takes the form

$$F(v) = \frac{1}{\sqrt{2\pi\sigma^2}} \int_0^\infty W(y)\exp\left[\frac{-(v - y)^2}{2\sigma^2}\right] dy \tag{20}$$

where σ^2, the variance of the spreading function, is independent of retention volume.

3.2.2. Nonuniform Gaussian Spreading Function

The integral equation takes the form

$$F(v) = \int_0^\infty W(y)\, \frac{1}{\sqrt{2\pi\sigma^2(y)}} \exp\left[\frac{-(v - y)^2}{2\sigma^2(y)}\right] dy \tag{21}$$

When peak broadening is not excessive, σ does not vary appreciably with retention volume; when the particle size distribution is relatively narrow, an excellent approximation to Equation 21 is given by

$$F(v) = \frac{1}{\sqrt{2\pi\sigma(v)^2}} \int_0^\infty W(y)\exp\left[\frac{-(v - y)^2}{2\sigma^2(v)}\right] dy \qquad (22)$$

Equation 22 is the basis for the derivation of correction equations for dispersion in the detector cell to be made later in this chapter. However, its range of applicability has not been examined in any detail.

3.2.3. General Spreading Function

Provder and Rosen (14) proposed the use of a general statistical shape function to account for deviations of the spreading function from the Gaussian shape. It has the form

$$G(x) = \phi(x) + \sum_{n=3}^\infty (-1)^n A_n \phi^n(x)/n! \qquad (23)$$

where

$$\phi(x) = \frac{1}{\sqrt{2\pi}} \exp\left(\frac{-x^2}{2}\right)$$

$$x = \frac{v}{\sigma}$$

and $\phi^n(x)$ denotes the nth-order derivative. The coefficients A_n are functions of μ_n, the nth-order moments about the mean retention volume, μ_1, of the normalized detector response for a single species. The first two coefficients are of direct statistical significance and also represent the most useful terms in the infinite series for applications in chromatography. Coefficients A_3 and A_4 are defined as:

$$A_3 = \frac{\mu_3}{\mu_2^{3/2}} \qquad (24)$$

$$A_4 = \frac{\mu_4}{\mu_2^2} - 3 \qquad (25)$$

where μ_2 is the variance and is an equivalent symbol to σ^2. The coefficient

A_3 provides an absolute statistical measure of skewness (when $\mu_3 = 0$, the spreading function is symmetrical about the mean retention volume μ_1 or y). When $\mu_3 > 0$, skewing is toward longer retention volumes. The coefficient A_4 provides a statistical measure of flattening or kurtosis. When $A_4 > 0$, the shape function is taller and slimmer than a Gaussian, and so forth. Tung and Runyon (15) used a simpler form to fit skewed detector responses in the size exclusion chromatography (SEC) of polymer molecules. Silebi (16) showed recently that skewed instrumental spreading functions derived from the plug-flow dispersion model (17) adequately fit data for particle separations by hydrodynamic chromatography (HDC). This spreading function has the form

$$G(v, y) = \frac{1}{2\sqrt{\pi \mathrm{Pe}^{-1}(v/y)}} \exp\left[\frac{-(v - y)^2}{4\mathrm{Pe}^{-1}v/y)}\right] \qquad (26)$$

where $\mathrm{Pe} = UL/D'$ is the Peclet number, U is the superficial velocity in the column, L is the length of the packed bed, and D' is a dispersion coefficient. The plug-flow dispersion model predicts symmetrical broadening in the packed bed; however, when dispersion is large, the detector gives a response that is skewed toward longer retention volumes. For small dispersion, $\mathrm{Pe} > 100$, $G(v, y)$ reduces to a Gaussian shape.

4. ANALYTICAL SOLUTION OF THE INTEGRAL EQUATION— PARTICLE DIAMETER AVERAGES

4.1. Introduction

The first solutions of this kind were based on the use of bilateral Laplace transformations and the integral equation with a uniform instrumental spreading function (13, 14) and were applied to SEC of polymer molecules. The first application to the chromatography of spherical suspensions was made by Hamielec and Singh (18), and Husain et al. (2). Yau et al. (19) obtained similar solutions for the case of a uniform spreading function and a linear molecular weight calibration curve (equivalent to a linear particle diameter-retention volume calibration curve in this context). Yau et al. (19) focused on dispersion in the detector cell, as did Hamielec et al. (20, 21) in accounting for a nonuniform Gaussian spreading function and a nonlinear calibration curve.

4.2. Uniform Spreading Function and Linear Particle Diameter Calibration Curve

4.2.1. Gaussian Spreading Function

For the special case of a turbidity detector in the Rayleigh scattering regime, correction equations for whole sample diameter averages are given by the following equations in which (c) refers to the corrected and (uc) to uncorrected average diameters (9, 18).

$$D_N(c) = D_N(uc) \exp[\tfrac{11}{2}(D_2\sigma)^2] \tag{27}$$

$$D_S(c) = D_S(uc) \exp[5(D_2\sigma)^2] \tag{28}$$

$$D_V(c) = D_V(uc) \exp[\tfrac{9}{2}(D_2\sigma)^2] \tag{29}$$

$$D_{SS}(c) = D_{SS}(uc) \exp[\tfrac{7}{2}(D_2\sigma)^2] \tag{30}$$

$$D_W(c) = D_W(uc) \exp[\tfrac{5}{2}(D_2\sigma)^2] \tag{31}$$

$$D_\tau(c) = D_\tau(uc) \exp[\tfrac{3}{2}(D_2\sigma)^2] \tag{32}$$

The linear particle diameter-retention volume calibration curve is given by

$$D(v) = D_1 \exp(-D_2 v) \tag{33}$$

where D_1 is the intercept and D_2 is the slope, with $D_1, D_2 > 0$.

When the more general, Mie scattering theory is applied, the use of Equation 19 to obtain correction equations similar to those given by Equations 27–32 cannot be used. One can, however, derive an analytical expression for the moments of the size distribution of the detector cell contents (22). This expression, along with $N(v, y)$, can be used to generate the diameter averages of the whole sample.

In contrast to the method based on Rayleigh scattering, the method based on Mie scattering theory (1) is not restricted to small particles, (2) may consider chemical absorption of light, and (3) can be extended to permit the use of a nonuniform Gaussian spreading function and a non-linear particle diameter-retention volume calibration curve (22). For practical purposes, it may be assumed that Rayleigh scattering is valid when $\alpha < 1$. With $\alpha = 1$ and $\lambda = 750$ nm, particles with diameters less than about 240 nm would experience Rayleigh scattering. Many latexes are in this size range.

4.2.2. General Spreading Function

For the case of skewed uniform instrumental spreading and a linear particle diameter-retention volume calibration curve, correction equations for a general detector of Type 1 follow (18)

$$D_N(\text{c}) = D_N(\text{uc}) \exp\left[\left(\gamma - \frac{1}{2}\right)(D_2\sigma)^2\right]\left(\frac{[1 + \alpha'\,(\gamma D_2)^3]}{\{1 - \alpha'[(1 - \gamma)D_2]^3\}}\right) \quad (34)$$

$$D_W(\text{c}) = D_W(\text{uc}) \exp\left[\left(\gamma - \frac{7}{2}\right)(D_2\sigma)^2\right]\left(\frac{\{1 - \alpha'[(3 - \gamma)D_2]^3\}}{\{1 - \alpha'[(4 - \gamma)D_2]^3\}}\right) \quad (35)$$

where α' is an adjustable parameter in the spreading function that accounts for skewing ($\alpha' = A_3\sigma^3/6$, $A_4 = 0$). When $\alpha' = 0$, $G(v, y)$ is a uniform Gaussian. For the case of a refractometer detector ($\gamma = 3$), and a turbidity detector in the Rayleigh scattering regime ($\gamma = 6$), Equations 34 and 35 reduce to

$$D_N(\text{c}) = D_N(\text{uc}) \exp\left[\frac{5}{2}(D_2\sigma)^2\right]\left[\frac{(1 + 27\alpha'D_2^3)}{(1 + 8\alpha'D_2^3)}\right] \quad (36)$$

$$D_W(\text{c}) = D_W(\text{uc}) \exp\left[-\frac{1}{2}(D_2\sigma)^2\right]\left[\frac{(1)}{(1 - \alpha'D_2^3)}\right] \quad (37)$$

for a refractometer detector, and

$$D_N(\text{c}) = D_N(\text{uc}) \exp\left[\frac{11}{2}(D_2\sigma)^2\right]\left[\frac{(1 + 216\alpha'D_2^3)}{(1 + 125\alpha'D_2^3)}\right] \quad (38)$$

$$D_W(\text{c}) = D_W(\text{uc}) \exp\left[\frac{5}{2}(D_2\sigma)^2\right]\left[\frac{(1 + 27\alpha'D_2^3)}{(1 + 8\alpha'D_2^3)}\right] \quad (39)$$

for the turbidity detector. Equations 36–39 are derived in the Appendix. It is of interest to compare the magnitude of the corrections for dispersion for these two detector types. The refractometer response requires a smaller correction for both D_N and D_W. The correction to D_N for the turbidity dectector is excessive and suggests that this type of detector is not suitable for the measurement of D_N in the Rayleigh scattering regime, because the extinction coefficient is proportional to diameter raised to the fourth power. Small changes in the detector response at the low-particle diameter end are magnified enormously when translated into a calculation of total number of particles. It is also interesting to note that D_W is bracketed by the two detector types, giving upper and lower bounds, and also that the skewing correction increases all of the diameter averages. The symmetric dispersion and skewing corrections to D_W tend to compensate, giving a small correction to D_W when the refractometer detector is employed. From the point of view of dispersion error in calculating particle diameter averages, the refractometer is superior to the turbidimeter. This is due to the more favorable dependence of detector signal on particle diameter and, in essence, is equivalent to the use of absorbing wavelengths to increase the extinction coefficient of smaller particles to give a more favorable dependence of detector signal on particle diameter.

4.3. Nonuniform Spreading Function and Nonlinear Particle Diameter-Retention Volume Calibration Curve

This case has been treated by Husain et al. (9, 22). A novel method for identifying and estimating the parameters of the instrumental spreading function for column chromatography was developed and applied to the SEC of particle suspensions (23). This revealed that, for SEC, the spreading function of polystyrene latex standards in the size range 85–312 nm are skewed toward longer retention volumes. The Provder and Rosen general spreading function gives reasonable fit to experimentally measured spreading functions for particles in the size range 85–220 nm. For the 312-nm standard, the fit was poor. It would be of interest to compare experimental $G(v, y)$ with the spreading function predicted by the plug-flow dispersion model. It appears that for HDC or SEC of particles, a skewed instrumental spreading function should be used to properly account for dispersion for particles greater than about 100 nm in diameter.

4.4. Application of Correction Equations for Dispersion

Many experimental and numerical examples are provided by Husain et al. (2, 9, 18, 22). A sequence of calculations which may be done on a raw detector response to provide the frequency distribution and diameter averages of the whole sample is given in the Appendix.

5. NUMERICAL SOLUTION OF THE INTEGRAL EQUATION—CORRECTED DETECTOR RESPONSE

Several numerical methods were reported for the solution of the integral equation. These were reviewed by Friis and Hamielec (24) and recently evaluated by Silebi and McHugh (5) for their application to particle chromatography. They concluded that the method of Ishige et al. (25) performs better than other available methods. A noteworthy undesirable feature of the method, however, is its tendency to overestimate the number of small particles in a polydispersed sample. Modifications of Ishige's algorithm fail to overcome this defect (26). Until a more effective numerical method, which applies universally, is developed for solving for the corrected detector response, $W(y)$, it is recommended that analytical methods described in the last section be considered along with numerical methods for calculating particle diameter averages.

6. DISPERSION IN THE DETECTOR CELL

6.1. Introduction

Early attempts to solve the integral equation focused on the detector responses (raw and dispersion corrected), instrumental spreading, and whole sample particle diameter averages. With the advent of sophisticated detector systems, a need arose to examine dispersion in the detector cell to properly interpret detector responses. The first attempt was made by Yau et al. (19), who considered a uniform Gaussian spreading function and a linear molecular weight-retention volume calibration curve in the SEC of polymer molecules. This solution was then generalized to include a nonuniform Gaussian spreading function and a nonlinear molecular weight calibration curve (20). More recently, an analytical solution of the integral equation for $W(y)$ was obtained using this generalized solution (21).

6.2. Size Distribution of Detector Cell Contents

An excellent approximation to $W(v, y)$ is given by

$$W(v, y) = [a(v)y^2 + b(v)y + c(v)]G(v, y) \qquad (40)$$

with

$$F(v) = \int_0^\infty [a(v)y^2 + b(v)y + c(v)]G(v, y)\, dy \qquad (41)$$

The parameters of $G(v, y)$ are presumed known from the usual calibration procedures, and this leaves two additional parameters among $a(v)$,

$b(v)$, and $c(v)$ that must be determined to find $W(v, y)$ and $W(y)$. The term $W(v)$ is given by

$$W(v) = \frac{W(v, v)}{G(v, v)} \tag{42}$$

or, approximately, by

$$W(v) = a(v)v^2 + b(v)v + c(v) \tag{43}$$

For the case of a nonuniform Gaussian instrumental function, and when dispersion is not excessive, $W(v, y)$ can be adequately approximated by a Gaussian distribution of the form (21)

$$W(v, y) = \frac{F(v)}{\sqrt{2\pi\bar{\sigma}(v)^2}} \exp\left\{\frac{-[y - \bar{y}(v)^2]}{2\bar{\sigma}(v)^2}\right\} \tag{44}$$

where $F(v)$ is a normalization factor.

6.3. Diameter Averages in the Detector Cell

Relationships for $D_N(v)$ and $D_W(v)$ are now derived. The frequency distribution in the detector cell (Type 2 detector) may be expressed as

$$f(D, v) \, dD = -\frac{[W(v, y)/K(y)D^2(y)] \, dy}{\displaystyle\int_0^\infty [W(v, y)/K(y)D^2(y)] \, dy} \tag{45}$$

and, therefore,

$$D_N(v) = \int_0^\infty Df(D, v) \, dD = \frac{\displaystyle\int_0^\infty [W(v, y)/K(y)D(y)] \, dy}{\displaystyle\int_0^\infty [W(v, y)/K(y)D^2(y)] \, dy} \tag{46}$$

With local linearization, a nonlinear particle diameter-retention volume calibration curve and the extinction coefficient can be expressed as

$$D(y) = D_1(v) \exp[-D_2(v)y] \tag{47}$$

where $D_1(v), D_2(v) > 0$, and

$$K(y) = K_1(v) \exp[-K_2(v)y] \tag{48}$$

where $K_1(v)$, $K_2(v) > 0$.

$$D_N(v) = D_1(v) = \frac{\int_0^\infty W(v, y) \exp[E(v)y]\, dy}{\int_0^\infty W(v, y) \exp[P(v)y]\, dy} \tag{49}$$

where $E(v) = D_2(v) + K_2(v)$ and $P(v) = 2D_2(v) + K_2(v)$.

For the special case where the instrumental spreading function is a nonuniform Gaussian distribution,

$$\int_0^\infty W(v, y) \exp[E(v)y]\, dy$$

$$= \exp[E(v)y]F[v + E(v)\sigma^2(v)] \exp\left\{\frac{[E(v)\sigma(v)]^2}{2}\right\} \tag{50}$$

and, therefore,

$$\frac{D_N(v)}{D(v)} = \frac{F[v + E(v)\sigma(v)^2]}{F[v + P(v)\sigma(v)^2]} \exp\left\{\frac{[E^2(v) - P^2(v)]\sigma(v)^2}{2}\right\} \tag{51}$$

In the Rayleigh scattering regime, Equation 51 reduces to

$$\frac{D_N(v)}{D(v)} = \frac{F[v + 5D_2(v)\sigma(v)^2]}{F[v + 6D_2(v)\sigma(v)^2]} \exp\left\{-\frac{11}{2}[D_2(v)\sigma(v)]^2\right\} \tag{52}$$

The equation for $D_W(v)$ in Mie scattering is of identical format to Equation 51, but now

$$E(v) = K_2(v) - 2D_2(v) \quad \text{and} \quad P(v) = K_2(v) - D_2(v)$$

In the Rayleigh scattering regime, the equation for D_W follows

$$\frac{D_W(v)}{D(v)} = \frac{F[v + 2D_2(v)\sigma(v)^2]}{F[v + 3D_2(v)\sigma(v)^2]} \exp\left\{-\frac{5}{2}[D_2(v)\sigma(v)]^2\right\} \tag{53}$$

The polydispersity $D_W(v)/D_N(v)$ in the detector cell may be found by simply dividing Equation 53 by Equation 52. Expressions such as Equations 51 and 52 can readily be found for other diameter averages. Equations of this type should find use with detectors based on turbidity–spectra and photon correlation spectroscopy.

7. ANALYTICAL SOLUTION OF THE INTEGRAL EQUATION—CORRECTED DETECTOR RESPONSE

7.1. Nonuniform Gaussian Spreading Function

For the case of nonuniform Gaussian spreading function analytical expressions for $D_N(v)$, $D_W(v)$, and other diameters, averages of the detector cell contents are available. One can then readily express $a(v)$, $b(v)$, and $c(v)$ of the analytical solution given by Equation 43 in terms of the known quantities $D_2(v)$ and $\sigma(v)^2$. For the limiting case of a nonuniform Gaussian spreading function and small levels of dispersion, an analytical solution based on Equation 44 was developed and applied to the SEC of polymer molecules.

7.2. General Spreading Function

When the instrumental spreading function is skewed, the mathematics for an analytical solution of the integral equation to give $W(y)$ has not been worked out, and one must either use a numerical solution or determine $W(y)$ instrumentally. The starting point for an instrumental method of determining $W(y)$ is Equation 40. At least two moments of $W(v, y)$ must be measured on-line to permit solutions for $a(v)$, b(v), and $c(v)$. Instrumental methods that, in principle, could provide such information on-line are discussed in the following section.

8. INSTRUMENTAL CORRECTION FOR DISPERSION

8.1. Introduction

An HDC or SEC operating with normal resolution should provide unimodal and relatively narrow frequency distributions of particle size in the detector cell across the chromatogram of a whole sample. Therefore, a detector system that can provide, say, two moments of the frequency distribution and the particle concentration of the detector cell contents should, in principle, provide a measure of the frequency distribution of the whole sample, and this measure should be largely independent of the resolution of the chromatograph.

To date, the use of a detector system to this end has not been reported. There are, however, at least two detector systems based on turbidity–spectra (27, 28) and quasi-elastic light scattering that seem to have the potential for this task. A discussion of these detectors follows.

8.2. Turbidity-Spectra

The turbidity in the detector cell for a polydispersed suspension can be expressed as

$$\tau(\lambda, v) = N(v) \int_0^\infty K(D)\pi D^2 f(D, v) \, dD \tag{54}$$

where $N(v)$ is the total number of particles per unit volume in the detector cell at retention volume v. It is assumed that $f(D, v)$ is unimodal with two adjustable parameters, giving in total four unknowns [including $N(v)$ and the refractive index of the particles]. In the Mie scattering regime, a minimum of four wavelengths must be employed to solve for these quantities. In the Rayleigh scattering regime, measurements of the turbidity at different wavelengths does not provide independent information. Thus, turbidity–spectra cannot be used to measure the frequency distribution and $N(v)$ in the detector cell. Husain et al. (27) evaluated the use of turbidity–spectra for this purpose, using a computer simulation. They found that representation of $f(D, v)$ with a log-normal distribution gave good results for most operational ranges of HDC and SEC. However, when resolution is poor, the size distribution in the detector cell can be bimodal, even though the whole sample is unimodal.

8.3. Photon Correlation Spectroscopy

From diffusion coefficient measurements, this technique can provide an indirect measure of particle size in the range 20–10,000 Å, without a need for particle refractive index and concentration (29–32, 34). (See Chapters 1, 3 and 4). Measurements near 180° require very short counting time (5–10 sec for 200-Å and 25–30 sec for 5000-Å particles) and do not require filtration of the sample (32). For polydispersed suspensions, one obtains approximately the z-average particle diameter and a relative polydispersity index (34). It therefore appears possible, in principle, to use a detector based on photon correlation spectroscopy in series with, say, a turbidimeter as a detector system for HDC and SEC to estimate $f(D, v)$, $N(v)$, and the refractive index of the particles. The use of photon correlation spectroscopy along with turbidimetry would permit one to work in the Rayleigh as well as in the Mie scattering regime and should make solving the parameters of $f(D, v)$ less ambiguous (33).

9. FUTURE DIRECTIONS FOR DETECTOR DEVELOPMENT

A serious limitation of HDC and SEC of particle suspensions is the extensive axial dispersion (peak broadening) that must be corrected to obtain absolute particle size distributions. Other limitations are briefly reviewed in Chapter 1. The large axial dispersion results in skewed instrumental spreading functions, compounding the problem of dispersion correction. This problem can be solved instrumentally with a detector system that can measure particle concentration and size distribution in the detector cell. It is therefore recommended that such a detector system, based on a combination of turbidity–spectra and photon correlation spectroscopy, be developed for application with HDC and SEC of particle suspensions. It should be noted that, as discussed in Chapter 7, axial dispersion is not a serious limitation to sedimentation field flow fractionation or to the use of the Joyce–Loebl disk centrifuge.

NOTATION

A	Absorbance, dimensionless.
A_n	Coefficient of nth-order moment, μ_n.
c	Concentration, g/cm^3.
D	Particle diameter, cm.
D_n	Number average diameter, nm.
D_W	Weight average diameter, nm.
D_S	Surface average diameter, nm.
D_{SS}	Specific surface average diameter, nm.
D_V	Volume average diameter, nm.
D_τ	Turbidity average diameter, nm.
$D(y)$	Diameter of particles with mean retention volume y, nm.
D_1	Intercept of ln particle diameter—retention volume calibration curve, nm.
D_2	Slope of ln particle diameter—retention volume calibration curve.
D'	Dispersion coefficient, cm^2/sec.

$F(v)$	General detector response at retention volume v.
$G(v, y)$	Spreading function or normalized detector response for a particle of diameter $D(y)$.
I	Intensity of emerging beam
I_0	Intensity of incident beam.
K	Extinction coefficient, dimensionless.
$K(y)$	Extinction coefficient for particles of diameter $D(y)$.
l	Optical path length, cm.
L	Length of packed bed, cm.
m	Ratio of refractive indices of particle and medium.
N	Particle concentration, number/cm^3.
$N(v)$	Total number of particles per unit volume in detector cell.
$N(v, y)\, dv\, dy$	Number of particles within the volume range of $v - (v + dv)$ with a mean retention volume of $y - (y + dy)$.
Pe	Peclet number, UL/D'.
U	Superficial velocity in column, cm/sec.
v	Retention volume, cm^3.
y or μ_1	Mean retention volume of a distribution of particles, cm^3.
$W(y)$	Detector response corrected for dispersion.
$W(y)\, dy$	Area under detector response owing to particles of diameter $D(y)$.
$W(v, y)\, dv\, dy$	Area under detector response in the retention volume range $v - (v + dv)$ resulting from particles with a mean retention volume of $y - (y + dy)$.
α	Dimensionless size parameter ($\alpha = \pi D/\lambda$). Defined as x in Chapters 1, 5, and 6.
α'	Adjustable parameter in the spreading function that takes into account skewing.
γ	Exponential particle response factor in which $\gamma = 3$ for refractive index and $\gamma = 6$ for turbidity in the Raleigh scattering regime.
η_m	Refractive index of medium.
η_s	Refractive index of dispersion.
λ	Wavelength of light, cm.
μ_n	The nth-order variance about the mean retention volume, μ, of normalized detector response for a single species. For $n = 2$, $\mu_2 = \sigma^2$.
ρ_P	Particle density, g/cm^3.

σ^2 Variance of spreading function, cm^6.

τ Turbidity, cm^{-1}.

REFERENCES

1. A. Husain, A. E. Hamielec, and J. Vlachopoulos, *J. Liquid Chromatogr.*, **4** (Suppl. 2), 295 (1981).
2. A. Husain, J. Vlachopoulos, and A. E. Hamielec, *J. Liquid Chromatogr.*, **2**, 193 (1979).
3. G. Mie, *Ann. Phys.*, **25**, 377 (1908).
4. W. Heller and W. J. Pargonis, *J. Chem. Phys.*, **26**, 498 (1957).
5. C. A. Silebi and A. J. McHugh, *J. Appl. Polym. Sci.*, **23**, 1699 (1979).
6. D. J. Nagy, Ph.D. Thesis, Lehigh University (1979).
7. D. J. Nagy, C. A. Silebi, and A. J. McHugh, in R. M. Fitch (Ed.), *Polymer Colloids II*, Plenum Press, New York, 1980.
8. W. Heller and R. M. Tabibian, *J. Colloid Sci.*, **12**, 25 (1957).
9. A. Husain, A. E. Hamielec, and J. Vlachopoulos, *ACS Symp. Ser.*, **138**, 47 (1980).
10. H. B. Zimm and W. B. Dandliker, *J. Phys. Chem.*, **58**, 644 (1954).
11. H. Coll and G. R. Fague, *J. Colloid Interface Sci.*, **76**, 116 (1980).
12. M. Nakagaki and W. Heller, *J. Appl. Phys.*, **27**, 9 (1956).
13. A. E. Hamielec and W. H. Ray, *J. Appl. Polym. Sci.*, **13**, 1319 (1969).
14. T. Provder and E. M. Rosen, *Sep. Sci.*, **5**, 437, 485 (1970).
15. L. H. Tung and F. R. Runyon, *J. Appl. Polym. Sci.*, **13**, 2397 (1969).
16. C. A. Silebi, FACSS Symposium, Abstract 212, Philadelphia, PA, 1981.
17. M. Hess and R. F. Kratz, *J. Polym. Sci.*, *A-2*, **4**, 731 (1966).
18. A. E. Hamielec and S. Singh, *J. Liq. Chromatogr.*, **1**, 187 (1978).
19. W. W. Yau, J. J. Kirkland, D. D. Bly, and H. J. Stoklosa, *J. Chromatogr.*, **125**, 219 (1976).
20. A. E. Hamielec, *J. Liq. Chromatogr.*, **3**, 381 (1980).
21. A. E. Hamielec, H. J. Ederer, and K. H. Ebert, *J. Liq. Chromatogr.*, **4**, 1697 (1981).
22. A. Husain, A. E. Hamielec, and J. Vlachopoulos, *J. Liq. Chromatogr.*, **4**, 425 (1981).
23. A. Husain, A. E. Hamielec, and J. Vlachopoulos, *J. Liq. Chromatogr.*, **4**, 459 (1981).
24. N. Friis and A. E. Hamielec, *Adv. Chromatogr.*, **13**, 41 (1975).
25. T. Ishige, S. I. Lee, and A. E. Hamielec, *J. Appl. Polym. Sci.*, **15**, 1607 (1971).

26. A. Husain, Ph.D. Thesis, McMaster University (1980).

27. A. Husain, J. Vlachopoulos, and A. E. Hamielec, *J. Liq. Chromatogr.*, **2,** 517 (1979).

28. A. E. Hamielec, *J. Liq. Chromatogr.*, **1,** 555 (1978).

29. B. Chu, *Ann. Rev. Phys. Chem.*, **21,** 145 (1970).

30. B. Chu, *Laser Light Scattering*, Academic Press, New York, 1974.

31. B. J. Berne and R. Pecora, *Dynamic Light Scattering*, Wiley, New York, 1976.

32. E. J. Derderian and T. B. MacRury, *J. Dispersion Sci. Technol.*, **2,** 345 (1981).

33. R. L. Zollers, *J. Dispersion Sci. Technol.*, **2,** 331 (1981).

34. J. C. Brown, P. N. Pusey, and R. Dietz, *J. Chem. Phys.*, **62,** 1136 (1975).

APPENDIX

I. Derivation of Correction Equations for a Skewed Spreading Function

A uniform and skewed spreading function based on the general spreading function (Equation 23) is given by

$$G(v, y) = \left(\frac{1 - A_3}{2\sigma}\right) \left[\frac{(v - y) - (v - y)^3}{3\sigma^2}\right]$$
$$\times \left(\frac{1}{\sqrt{2\pi\sigma^2}}\right) \exp\left[\frac{-(v - y)^2}{2\sigma^2}\right] \quad (I.1)$$

with σ and A_3 independent of the retention volume. The bilateral LaPlace transformation of $G(v, y)$ is given by

$$\overline{G}(S) = \frac{1 - A_3 S^3 \sigma^3}{6} \exp\left(\frac{S^2 \sigma^2}{2}\right) \quad (I.2)$$

Correction equations for a light scattering photometer detector in the Rayleigh scattering regime ($\gamma = 6$) are now derived. The detector response is proportional to the number of particles in the detector cell times the particle diameter to the sixth power and, as a consequence, the weight-average particle diameter, D_W, is given by

$$D_W(\text{uc}) = \frac{\displaystyle\int_{-\infty}^{\infty} F(v)\, D^{-2}(v)\, dv}{\displaystyle\int_{-\infty}^{\infty} F(v)\, D^{-3}(v)\, dv} \quad (I.3)$$

Other particle diameter averages may be similarly expressed. We next assume that the particle diameter-retention volume calibration curve is linear on a semi-log plot and is given by

$$D(v) = D_1 \exp(-D_2 v) \tag{I.4}$$

with D_1, $D_2 > 0$ and constant. Substituting Equation I.4 into I.3, it is clear that

$$D_W(\text{uc}) = \frac{D_1 \overline{F}(-2D_2)}{\overline{F}(-3D_2)} \tag{I.5}$$

In a similar manner,

$$D_W(\text{c}) = \frac{D_1 \overline{W}(-2D_2)}{\overline{W}(-3D_2)} \tag{I.6}$$

Application of Equation 19 gives

$$\frac{D_W(\text{c})}{D_W(\text{uc})} = \frac{\overline{G}(-3D_2)}{\overline{G}(-2D_2)} \tag{I.7}$$

Substituting Equation I.2 into Equation I.7 gives

$$\frac{D_W(\text{c})}{D_W(\text{uc})} = \exp\left[\frac{5(D_2\sigma)^2}{2}\right]\left[\frac{(1 + 27\,D_2^3\sigma^3 A_3/6)}{(1 + 8\,D_2^3\sigma^3 A_3/6)}\right] \tag{I.8}$$

Equation I.8 is identical to Equation 39 when α' is set equal to $A_3\sigma^3/6$. Correction equations for other diameter averages and for other detector types can be derived in a similar manner.

II.   Calculation Sequences for Frequency Distribution and Particle Diameter Averages Starting with a Raw Detector Response (see Figure 1).

The use of HDC or SEC to measure the frequency distribution of particle diameter and averages of an unknown sample requires that the chromatograph first be calibrated to provide the particle diameter calibration curve $D(y)$ [or $D(v)$] and parameters of the spreading function (σ^2 and A_3). The use of $A_4 \neq 0$ is not recommended at this time. The particle diameter calibration curve can be obtained using narrow particle size distribution (PSD) standards. The spreading function parameters σ^2 and

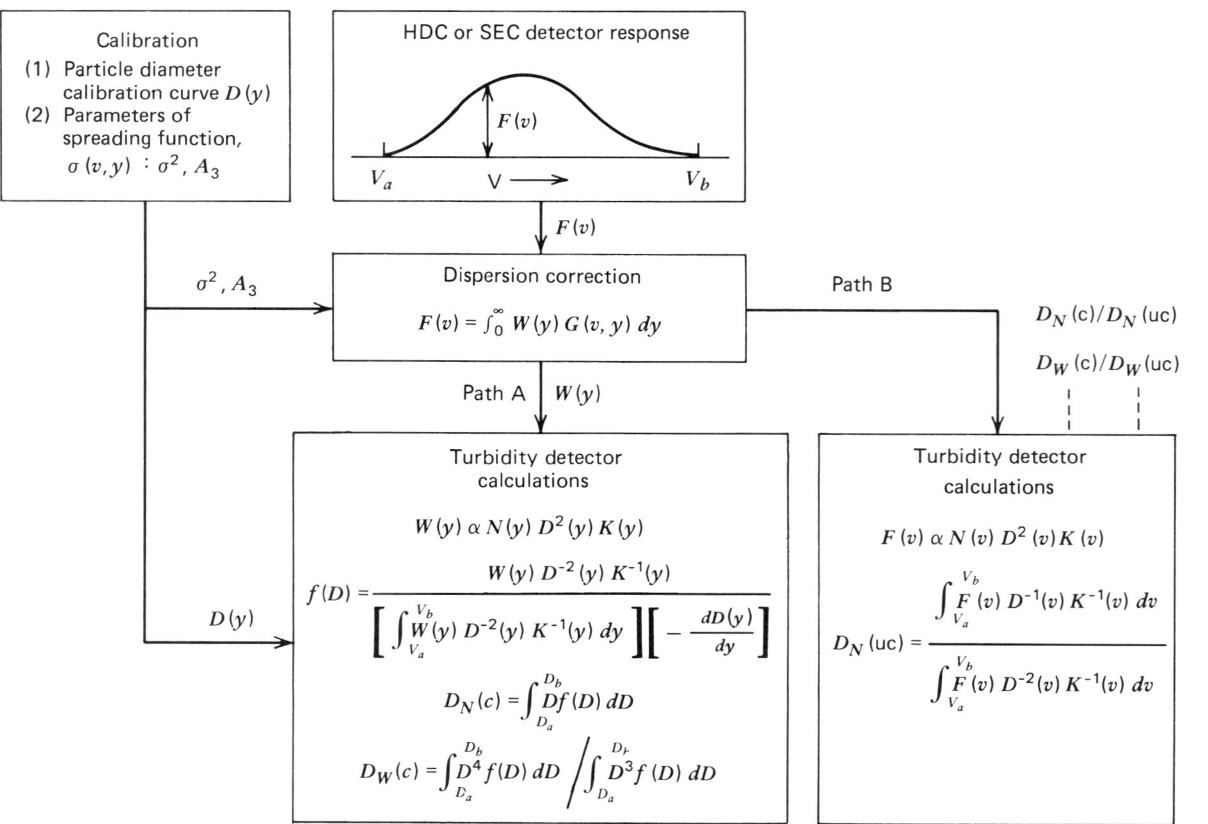

Figure 1. Two basic approaches of correcting for axial dispersion; D_a is the particle diameter, with mean retention volume V_a.

A_3 should be found using the method developed by Husain et al. (23). An alternative method is to apply Equations 36–39, when at least two diameter averages of the narrow PSD standards are known. This permits solutions for σ^2 and A_3. Equations 36–39 strictly apply for scattering in the Rayleigh regime; however, for narrow PSD, the change of extinction coefficient owing to diameter changes over the detector response is small and largely cancels out (9).

The first step in treating the raw detector response is to find the baseline and integration limits. Now, one has the choice of using one of two basic approaches of correcting for axial dispersion. One can either solve the integral equation numerically or analytically for the corrected response, $W(y)$, or use analytical solutions to correct the particle diameter averages directly, when applicable. The most common approach is to use a numerical solution to generate $W(y)$, followed by calculations of the frequency distribution and diameter averages. This is shown as Path A in Figure 1. The latter approach is shown in Path B.

CHAPTER

9

HYDRODYNAMIC CHROMATOGRAPHY OF HIGH MOLECULAR WEIGHT WATER-SOLUBLE POLYMERS

DAVID A. HOAGLAND, KAREN A. LARSON, and ROBERT K. PRUD'HOMME

Department of Chemical Engineering
Princeton University
Princeton, New Jersey

1. INTRODUCTION

Very high molecular weight polymers have found application in an increasing array of industries from cosmetics to oil production, where

277

knowledge of their molecular weight distribution could contribute significantly to the design of processes involving these materials. Commonly used polymers fitting into this category are xanthan gum, polyacrylamide, dextran, and the many derivatives of cellulose, such as carboxymethylcellulose, hydroxyethylcellulose, and so on. Particular note should be made of the diverse properties of the polymers on the list—some are polyelectrolytes and others are not. In addition, conformation varies from rigid rod-like to flexible coil. In each case, however, the hydrodynamic volume of an individual molecule is large, falling within the colloidal range. This chapter outlines a relatively inexpensive and rapid chromatographic technique for analyzing polydisperse samples of polymers with sizes in the range of 0.03–1.0 μm.

Several recent articles have discussed the successes and failures of conventional chromatography when applied to these polymers. Shawki and Hamielec (1) attempted to use size exclusion chromatography (SEC) packing materials with mean pore diameters as large as 1.4 μm to resolve partially hydrolyzed polyacrylamides with average molecular weights of 3–10 million. They concluded that the method was unsatisfactory, owing to exclusion of the high molecular weight fraction from the pores of the packing. The exclusion is indicated by the identical elution volume of the leading, or high molecular weight, edge of the chromatogram for all samples. A polydispersity calculated from such data will obviously be underestimated. A recent study of xanthan by Lambert et al. (2) under similar conditions may suffer from the same flaw. The reported polydispersity of 1.2 seems low, and the leading edge of the chromatograms for both native and hydrolyzed samples are identical.

Hydrodynamic chromatography (HDC), a technique developed to analyze latex particles similar in size to these ultra-high molecular weight polymers, provides a relatively simple alternative to SEC methods. (See Chapters 1 and 2 for brief reviews.) The support in an HDC column is a packed bed of impenetrable spheres with diameters of 10–50 μm. The velocity gradients created by the flow of mobile phase through the interstitial volume make separation possible. Small particles, under the influence of Browian motion, will sample regions of low fluid velocity near the packing bead surfaces. Since these regions are inaccessible to larger particles, the smaller particles will, on average, travel at a slightly lower velocity than the larger particles. An idealized separation of two spherical particles of different sizes in a capillary is shown in Figure 1. The ratio of average particle velocity to average fluid velocity, which defines the retention factor R_f, will typically range from 1.01–1.15 in an actual packed bed of spheres.

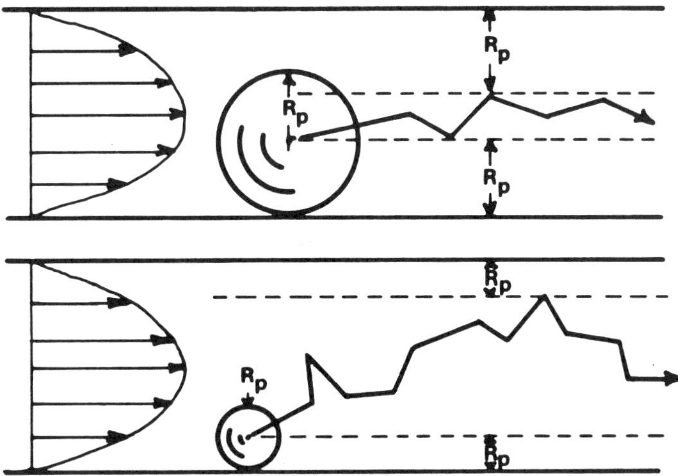

Figure 1. Capillary model of hydrodynamic chromatography separation. The average velocity of the larger particle, confined to the center of the tube, will be greater than the smaller particle, which can sample the velocity field near the wall. Brownian motion assures that the particles sample all radial positions with equal probability. R_p is the particle radius.

Most of the literature concerned with HDC is directed toward a better understanding of the resolution of samples of spherical latex particles. In addition to hydrodynamic exclusion, the electrostatic forces between latex particles and the support must be considered. Theoretical calculations based on both effects have been remarkably successful in describing latex particle separation (3–5). Recently, McGowan and Langhorst (6) described improved experimental methods and data analysis. Using more efficient packing techniques than previous investigators, delayed marker injection, and computer data acquisition and analysis, they accurately sized monodisperse populations of particles with diameters of 0.03–1.1 μm. In addition, they were able to accurately determine particle size distributions from known mixtures of these monodisperse populations.

Before considering HDC for polymer separation, it is necessary to estimate the effective size of the polymer in dilute solution to determine whether this value falls within the range of HDC resolution. Using the zero shear rate intrinsic viscosity and the average molecular weight, one can substitute into the following equation to obtain an estimate of the apparent hydrodynamic volume of a polymer molecule (7):

$$\text{vol} = \frac{0.40[\eta]M}{N_{Av}} \tag{1}$$

where $[\eta]$ is the intrinsic viscosity, M is the molecular weight, and N_{Av} is Avogadro's number. Xanthan, for example, has an intrinsic viscosity of about 5000 ml/g and a molecular weight of about 2 million. With these values, the calculated hydrodynamic volume, 0.007 μm^3, corresponds to an equivalent sphere diameter of 0.25 μm. Filtration through membranes of well-defined pore geometry yields a similar value for the size of xanthan, about 0.4 μm (8). Both values fall within the range of particle size for which HDC has proven most effective. The extension of HDC to polymer separation seems a logical step.

There are some important differences, however, between the use of HDC for separation of latex particles and its potential use in resolving polymer samples. Latex particles are rigid spheres, whereas polymer molecules are often deformable and nonspherical. The relationship between the measured hydrodynamic volume and polymer conformation during the chromatographic run may be a complicated function of flow rate, ionic strength, and surface charge on the column packing. Combined with the poorly defined bed geometry, these complexities make the separations more difficult to theoretically analyze than the separation of latex particles. The need for a mixed theoretical and experimental approach is obvious.

An additional problem arises from the large molecular size and enhanced solution viscosity of high molecular weight polymers; injected samples must be very dilute to inhibit both viscous fingering in the column and polymer–polymer interactions. Detection at low polymer concentrations is often difficult using conventional refractive index and UV-visible absorption detector systems. Fortunately, many high molecular weight polymers contain repetitive functional groups along the polymer chain. By using these groups in fluorescent tagging reactions, the resulting polymer can be detected at extremely low concentrations (often less than 50 ppm) using a fluorometer. An additional benefit of this approach is obtained if the tag attaches randomly along the chain, as is frequently observed. This results in a linear detector response with polymer mass concentration.

2. MATERIALS

2.1. Xanthan Gum

Production of xanthan gum (referred to here as xanthan), an extracellular microbial polysaccharide, occurs in a fermentation broth containing the bacterium *xanthamonas campestris*. Purification of xanthan requires isolation of the polymer from bacterial cell debris and other extraneous components of the broth. Xanthan microgels can obscure the properties of otherwise purified solutions and should also be removed. To handle these difficulties, several purification techniques are described in the literature (9–11). Xanthan samples used in this study were purified in two ways: early samples were treated according to the procedure of Holzwarth (12); later samples were purified by variants of Wellington's technique (13, 14). The final solutions prepared by the two methods show no obvious differences, although the latter method is the simpler of the two. Following Wellington's procedure, a 1000-ppm xanthan solution is prepared from Xanflood (Kelco Inc.) in 100 ppm of sodium chloride. Alcalase enzyme (100 ppm) obtained from Novo Laboratories is added to degrade cell bodies; this process is accelerated by elevating the solution temperature to 50°C for 1 hr. After this treatment, the solution is filtered through a 5.0-μm Millipore filter and dialyzed against 0.002 M sodium chloride, the solvent for the subsequent fluorescent tagging reaction. Dialysis continues for 3 days with frequent changes of dialysis fluid.

Holzwarth (15) developed a simple fluorescent tagging procedure for xanthan that involves the attachment of a fluorescein derivative to carboxyl groups found along the polymer chain. The fluorescence intensity of the resulting material varies with reactant concentration and reaction time. For xanthan, Holzwarth's procedure is followed, except for an extended (12–15 hr) reaction period. (Note: Holzwarth's paper should include the addition of 30 μl cyclohexyl isocyanide at the step where acetyldehyde is added.) After tagging, xanthan is precipitated three times with ethanol from 1.0% sodium chloride solutions. The purified material is then dialyzed against mobile phase for 3 days. The mobile phase used in this study is 0.002 M disodium hydrogen phosphate (Na_2HPO_4), 0.2% Brij 35 (ICI), 0.05% sodium lauryl sulfate, and 0.002 M sodium azide. Brij 35 is a nonionic surfactant composed of polyoxyethylene lauryl ether. This mobile phase has an ionic strength of about 0.01 M. Other eluants with higher ionic strengths were used successfully in the study of xanthan separations.

2.2. Partially Hydrolyzed Polyacrylamides

Table 1 lists the four different molecular weight polyacrylamides used in the HDC experiments. Since the partially hydrolyzed polyacrylamides contain carboxyl groups along the polymer chain, the procedure used for fluorescently tagging the polyacrylamides is nearly the same as the one used for tagging xanthan. Certain features of the tagging reaction, however, have been modified to give a more fluorescent product. These

Table 1. Polyacrylamides Used in This Study

Name	Abbreviation	MW^a	Manufacturer
Superfloc 208	SF208	16–18 million	American Cyanamid Co.
Superfloc 214	SF214	13–15 million	American Cyanamid Co.
Superfloc 204	SF204	2–4 million	American Cyanamid Co.
Polyacrylamide	Aldrich PA	200,000	Aldrich Chemical Co.

^{a} Nominal molecular weights and carboxyl substitution supplied by the manufacturers. The carboxyl substitution of SF214 has been reported as 38% [D. W. Rogers and G. W. Toling, *Can. Inst. Min. Bull.*, **71**, 152–158 (1978)]; it was assumed that the other polyacrylamides were in this range.

changes are briefly described here. Between 0.5–1.0 g of polymer is dissolved in approximately 150 ml of 1.0 M sodium chloride. The very high molecular weight polyacrylamides (i.e., SF208 and SF214) require an additional 30–40 ml of salt solution for complete dissolution. Once the material has dissolved, which requires approximately 24 hr of stirring with a magnetic stirrer, the tagging reaction is initiated with the addition of the following chemicals: 100 ml of 2:1 water/dimethyl sulfoxide, 0.05 g of 5-amino fluorescein dissolved in 3 ml of dimethyl sulfoxide, 68 μl of cyclohexyl isocyanide, and 68 μl of acetaldehyde. The reaction proceeds for 12 hr at room temperature and is followed by precipitation with 500 ml of isopropanol. The precipitate is then redissolved and dialyzed against distilled water for 3 days. After the addition of sodium azide to prevent microbial growth, the polymer is stored in a refrigerator.

In the initial polyacrylamide experiments, a much higher ionic strength mobile phase than the one used in the xanthan experiments was chosen to reduce sample viscosity by collapsing the polymer coils. The mobile phase, which had an ionic strength of about 0.31 M, consisted of 0.2% sodium lauryl sulfate, 0.10 M sodium sulfate, and 0.002 M sodium azide. Subsequent experiments showed that the ionic strength of the mobile phase had no significant effect on the retention factor, as will be discussed.

2.3. Dextran

Dextran of nominal 2 million molecular weight was obtained from Pharmacia Fine Chemicals and was tagged using the methods of de Belder and Granath (16). Since dextran does not contain carboxyl groups, the previous tagging reaction is not possible. Fluorescein isothiocyanate is attached to hydroxyl groups on dextran using a dibutyltin dilaurate catalyzed reaction. After tagging, the polymer is purified by several precip-

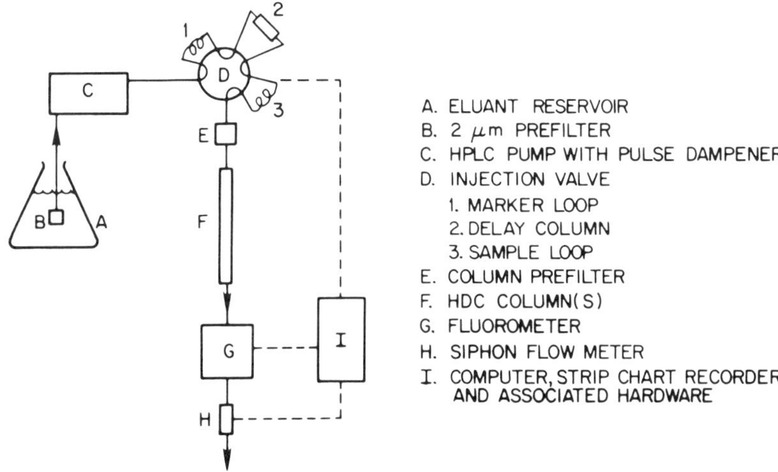

Figure 2. HDC apparatus. Note the computer interfacing for valve control and data acquisition.

itations with ethanol and dialyzed against the mobile phase, which consisted of 0.002 M disodium hydrogen phosphate, 0.2% Brij 35, 0.05% sodium lauryl sulfate, and 0.002 M sodium azide.

2.4. Latex Particle Standards

Latex particle standards were obtained from two commercial sources: 0.25- and 0.57-μm particles were purchased from Polysciences, Inc., and 0.176-, 0.198-, 0.305-, and 0.364-μm particles were purchased from Duke Scientific. In addition, H. Small of Dow Chemical Company provided us with 0.085- and 0.038-μm standard particles.

3. EXPERIMENTAL

The experimental apparatus is shown schematically in Figure 2. The mobile phase is delivered using a Milton Roy 396 Minipump with pulse dampener. To remove particulates, the mobile phase is filtered twice: once with a 0.22-μm Millipore filter during preparation, and again with a 2-μm frit located at the pump inlet. Maxima in the excitation and emission spectra of fluorescein occur at 494 and 515 nm, respectively (17). Consequently, a Kratos Model FS970 Spectrofluoro Monitor, possessing a 5-

μl flow cell, is operated at an excitation wavelength of 485 nm with a 500-nm emission filter. Both a Houston Instruments Series 4900 strip chart recorder and a Data General microNova computer store the data received from the fluorometer. The computer also controls the injection valve and uses signals from the siphon flowmeter to calculate the average flow rate during the run.

The overlap of sample and marker peaks is often encountered with species of small hydrodynamic size or when resolution is poor. A delayed marker injection technique made possible by a 12-port injection valve (Valco Instruments) eliminates the overlap problem; a small delay column is fitted between the sample loop and the marker loop. The delay column, with only 20% of the volume of the main HDC column, provides a time interval between marker injection and sample injection. Both injection loops are 20 μl. The volume of the delay element is determined by loading a fluorescein marker in both loops and measuring the interval between the resulting peaks as they pass the fluorometer. A 2-μm Rheodyne pre-column filter, placed between the injection valve and the HDC columns, insures the removal of any contaminants before the mobile phase enters the column. All connections in the system are made with stainless steel tubing, 1.6 mm O.D. and 0.254 mm I.D.

Xanthan and dextran separations are obtained using two columns, 9 mm I.D. × 20 cm, placed in series; each column contains monodisperse, 15-μm diameter, cross-linked poly(styrene-divinylbenzene) spheres (Dynospheres, Dow Diagnostics). These columns were slurry packed by M. Langhorst of Dow Chemical Company. The second system, used for polyacrylamide experiments, is comprised of a single column, 9 mm I.D. × 50 cm, filled with glass spheres obtained from Potter's Industries. This material was fractionated to approximately 18 μm (±4 μm) by elutriation. The column was then slurry packed following standard protocols (18–20). Since a high flow-rate, constant-pressure pump, commonly used in these procedures, was not available, a 900-ml syringe pump operating at 1200 psi was constructed from a high-pressure, double-acting hydraulic cylinder. After several trials, a slurry solvent consisting of a 1:4 mixture of glycerol and water was found most satisfactory.

Column efficiency was calculated using

$$N = 16 \left(\frac{t_r}{w}\right)^2 \tag{2}$$

where N is the number of theoretical plates generated in the column, t_r is the retention time of the marker (fluorescein) peak maximum, and w is the baseline width of the marker peak. The efficiency of the column packed with glass spheres was approximately 5000 plates/m at a flow rate of 1.4 ml/min as compared to 21,000 plates/m for the columns packed with Dynospheres at a flow rate of 3.5 ml/min. The lower plate count for the glass sphere column is likely due to the wider size distribution of the glass spheres used as packing and a less efficient slurry packing technique. Column efficiencies based on the latex particle standards are lower than those calculated with a low molecular weight marker. At 3.5 ml/min, the number of plates in the Dynosphere-packed column decreased by 30% for 0.078-μm particles and by 75% for 0.39 μm-particles.

4. RESULTS AND DISCUSSION

4.1. Xanthan Gum

4.1.1. Background

Our initial study of ultrahigh molecular weight polymers using hydrodynamic chromatography centered on the biopolymer xanthan (21). The currently accepted molecular structure of xanthan is depicted in Figure 3. The selection of this macromolecule for initial study was based on widespread commercial interest in the polymer, on current uncertainty regarding its size, and on its unusual and favorable molecular properties. These properties arise from the tendency of xanthan to adopt an extremely rigid and fairly rod-like conformation in solution (10, 22, 23). For comparison, recent measurements of the stiffness of xanthan (24) show it to be several times more extended than DNA, which has a persistence length of 0.06–0.09 μm (25, 26). Xanthan retains its structure under a wide range of temperatures, shear rates, and solution ionic strengths (12, 23). Thus, as the polymer flows through the HDC column, we assume that the conformation remains relatively unchanged by external forces. In contrast, the other two polymers examined in this study, polyacrylamide and dextran, are flexible and most likely deform under flow.

Although interest in xanthan has grown rapidly in recent years, details about its molecular weight and molecular weight distribution remain uncertain. Light-scattering and viscosity studies of the polymer indicate an average molecular weight of between 1.5 and 4.0 million (10, 27, 28). An electron microscopy study assigns a similar value to the molecular weight, about 2 million (13). Some researchers, however, reported values as high as 50 million (29). Although these measurements are admittedly difficult to make, the discrepancy among the different reported molecular weights is disconcerting, considering the relatively well-defined molecular weights

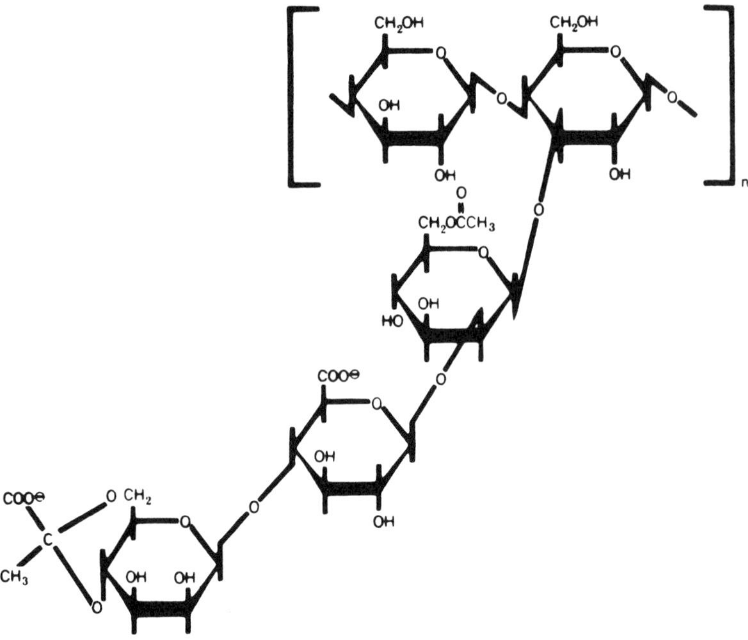

Figure 3. Repeat unit of xanthan gum.

one can obtain for smaller polymers. The discrepancy may be caused by differences in sample preparation and the existence of aggregates. Lack of an appropriate chromatographic analysis makes knowledge of the molecular weight distribution even more scanty than knowledge of the average molecular weight.

4.1.2. Xanthan Separation

Figure 4 displays a series of HDC separations on samples of native and sonicated xanthan. The chromatograms were made using an injection system in which sample and marker are mixed together. Operating conditions during this initial xanthan study differed from the description in the experiment section; the earlier xanthan runs were completed on a system similar to the one used for polyacrylamide studies (21). Data in the literature demonstrate that native xanthan degrades during sonication, and its effective size decreases accordingly (15). The series of chromatograms displayed in Figure 4 shows the gradual breakdown during degradation induced by an Ultrasonics Model W-370 Dismembrator. The apparatus

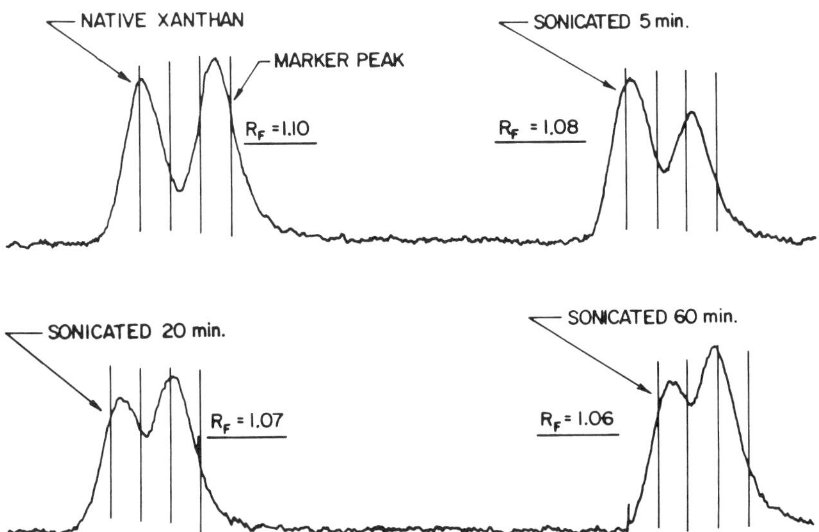

Figure 4. HDC of tagged xanthan and marker. Increasing sonication leads to lower molecular size, and this trend is reflected in the decreased separation of the sample and marker peaks (fluorescein). Column packing is 27-μm Zipax spherical silica beads and the mobile phase consists of 0.05 M sodium sulfate, 0.290% sodium lauryl sulfate, and 2 mM sodium azide (from reference 21).

was adjusted to supply 220 W to a 150-ml sample of polymer solution immersed in an ice bath. Although resolution is poor using this older column set, the chromatograms demonstrate that HDC can separate xanthan based on size. Excessive peak broadening arising from axial dispersion hinders calculation of polymer size distribution; fortunately, the dispersion can be reduced by improved column packing and injection. An example of a chromatogram of xanthan taken with the modified apparatus described in the experimental section is shown in Figure 5. The maximum in the xanthan peak occurs at an R_f of 1.062. To give meaning to this value, a calibration was made using spherical latex particles with diameters from 0.038 to 0.57 μm. The polystyrene latex particles were detected using their strong fluorescence in the ultraviolet region (excitation at 285 nm, detection above 320 nm). Salicylic acid was employed as a marker in these runs. From the calibration curve shown in Figure 6, it is possible to correlate the maximum in the xanthan peak to an effective spherical

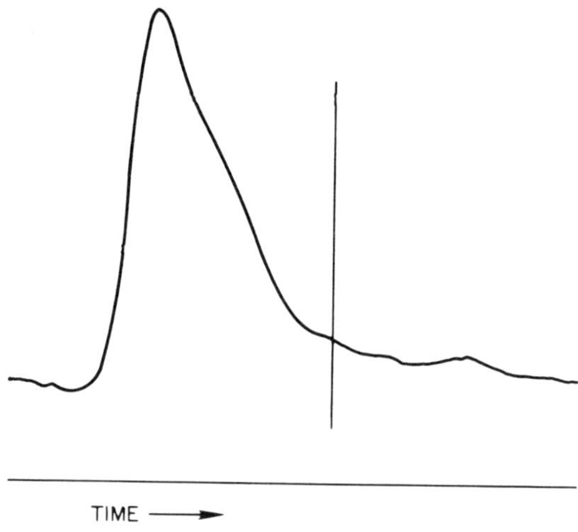

Figure 5.   Xanthan separation on modified HDC apparatus. The 50-ppm xanthan sample displays a broad peak and a large low molecular weight tail. The vertical line indicates the marker (low molecular weight) elution position. The chromatogram was obtained at a mobile phase flow rate of 1.6 ml/min on the system described in the experiment section. The fluorometer settings are 2.35 (adjust) and 0.002 (range); the strip chart recorder was set to 1 mV full scale at 4 cm/min.

size. An R_f of 1.062 translates into an equivalent sphere of diameter 0.35 μm. The trailing edge of the xanthan peak probably indicates low molecular weight material (and not dispersion in the column, as can be seen by comparing this chromatogram to the chromatogram of the essentially monodisperse 0.57-μm latex standard shown in Figure 7).

The eventual goal of analysis of macromolecules by HDC is rapid measurement of the molecular weight distribution of injected samples. Although Figure 5 may be qualitatively useful in characterizing xanthan, quantitative analysis based on experimental peaks is not yet justified. To more fully understand xanthan separation, a consideration of the flow of rod-like particles through a packed bed of spheres has been undertaken. These calculations were successful at qualitatively predicting xanthan peak location (30). The analysis assumes the polymer is partially oriented along streamlines as it travels through constrictions in the bed structure. With all orientations equally probable, the hydrodynamic radius would be half the polymer length. Alignment from flow, however, reduces the effective radius to less than this value. The calculation assumes that the effective size of the xanthan molecule is determined by a balance of rotational Brownian and hydrodynamic forces. Applying polymer kinetic

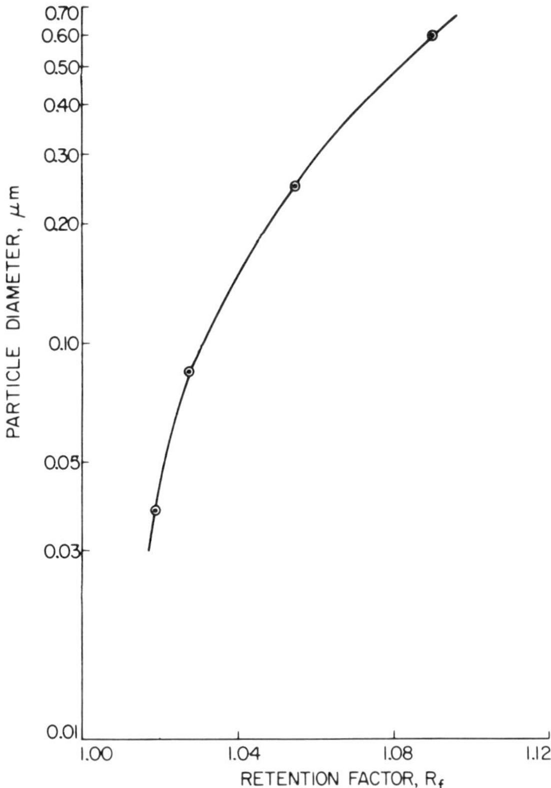

Figure 6. Calibration curve of column containing 15-μm Dynospheres. Calibration was obtained using spherical latex standards with a mobile phase ionic strength of 0.01 M. The flow rate was 1.6 ml/min for all samples. Injected solutions were in the range 0.1–0.5% solids. The excitation wavelength was 280 nm and a 320-nm emission filter was employed.

theory, the balance is used to calculate the orientational distribution function for the flowing polymer. In the range of flow rates examined, 0.30–1.70 ml/min, the calculations predict substantial orientation. Shear rates in the bed, for example, are on the order of 1000 s^{-1}, which is in the range where shear thinning in viscometric experiments is pronounced.

As a consequence of orientation by flow, the kinetic theory calculations predict a relationship between effective polymer size and flow rate. For a rod-like polymer, the R_f should decrease with increasing flow rate. The expected dependence was not observed over the range of flow rates studied. Refined calculations based on a more complete model may explain the discrepancy. Factors not previously considered, but likely to be im-

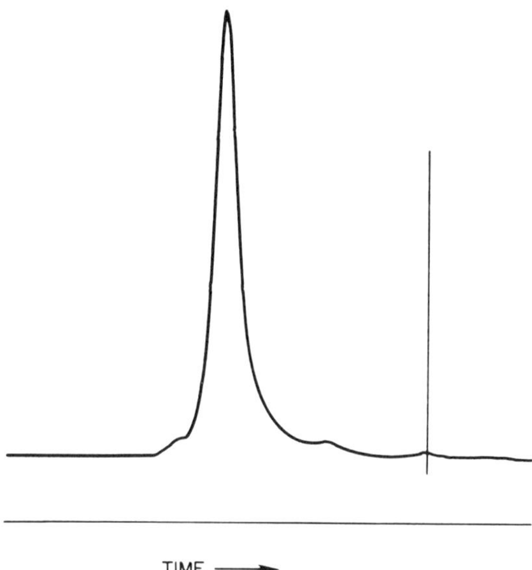

Figure 7. Chromatogram of 0.57-μm monodisperse latex spheres. This chromatogram is typical of the ones used to construct the calibration curve in Figure 6. The concentration of solids and the strip chart recorder were adjusted so that a comparison to Figure 5 can be made.

portant, could be polymer flexibility and time-dependent rotational motions arising from the periodicity of the flow channels.

4.2.   Partially Hydrolyzed Polyacrylamide

4.2.1.   Background

Polyacrylamides are an interesting class of polymers used as flocculants in wastewater treatment and as viscosity enhancers in tertiary oil recovery. As mentioned in Section 1, the motivation for the use of HDC for polyacrylamides has been the inability to characterize these high molecular weight water-soluble polymers using SEC. Unlike xanthan, polyacrylamides assume a flexible Gaussian coil conformation in solution. This structure makes the analysis of polyacrylamide separation more complicated than for xanthan, since the effective size of the polymer no longer depends on orientation by flow alone. Because of the elongational stresses arising from the converging and diverging flow in the bed, the polymer may undergo deformations that will also affect its measured size.

Table 2.   HDC Characterization of Polyacrylamides

Type	MW	R_f	Equivalent Diameter[a] (µm)
Aldrich PA	200,000	1.013	—
SF204	2–4 million	1.023	0.07
SF214	12–15 million	1.028	0.14
SF208	16–18 million	1.029	0.17

[a] Based on a polystyrene latex particle calibration curve.

4.2.2. *Polyacrylamide Separations*

After measurement of the retention factors of the four polyacrylamides, their equivalent particle diameters were found by interpolation from the calibration curve obtained from latex particles. The results are summarized in Table 2. Although there is a definite separation of the Aldrich PA from the fluorescein marker, the R_f of this sample is less than the R_f of any of the available latex standards; therefore, an extrapolation to determine its equivalent diameter was not attempted. The lower limit for resolution with the glass-packed column is apparently reached with a molecular weight of approximately 200,000. The R_f values of SF208 and SF214 are not significantly different; the lack of resolution between the peaks is probably due to low column efficiency. The results in Table 2, however, show that HDC is capable of distinguishing polyacrylamides with a 2-order of magnitude range in molecular weight. Further investigation using a more efficient column should yield better separations.

It was determined that R_f is not dependent on polymer concentration by diluting a 200-ppm solution to 125 ppm and comparing the R_f values of the two samples. The overlap concentration, the concentration at which interaction between polymer chains is expected to be significant, can be calculated from (31)

$$c^* = \frac{0.7}{[\eta]} \tag{3}$$

where c^* is the overlap concentration and $[\eta]$ is the intrinsic viscosity. The overlap concentration for SF208, 230 ppm, is above the concentration of samples injected during HDC experiments. Samples of SF208 tagged in separate batches yield the same retention factor, indicating the good reproducibility of HDC separations.

Table 3. A Comparison of Intrinsic Viscosity with R_f for Sonicated Samples of SF208

Sonication Time (sec)	$[\eta]$ (ml/g)	R_f
0	3070	1.052
10	2580	1.052
20	2015	1.032
50	760	1.027
100	—	1.016

4.2.3. Polyacrylamide Degradation

The ability of HDC to resolve polyacrylamides based on molecular weight is demonstrated by a second experiment in which the polymer SF208 is degraded by sonication. During this experiment, the polymer is sonicated for increasing lengths of time and is then characterized with intrinsic viscosity and R_f measurements. The intrinsic viscosities are determined using a four-bulb Cannon–Ubbelhode viscometer. Viscosities are extrapolated to zero shear rate, and the zero shear rate viscosities are then extrapolated to zero concentration. The intrinsic viscosity data are presented in Table 3, and the progression of degradation is illustrated in the HDC chromatograms of Figure 8. For reference, Figure 8(*a*) shows the chromatogram resulting from the injection of fluorescein loaded in both sample and marker loops. There is a considerable decrease in the intrinsic viscosity of the sample sonicated for 10 sec, but there is no change in R_f. Figures 8(*b*)(*c*), however, illustrate the drastic changes in peak shape upon degradation. In particular, a shoulder forms on the low molecular weight side of the peak in Figure 8(*c*). An effective R_f is determined by drawing lines tangent to the sides of the sample peak and defining the intersection point of these lines as the average R_f. The procedure is straightforward for symmetric peaks but is ambiguous when chromatograms show obvious bimodal size distributions, as in Figures 8(*c*) and (*d*). The observed differences in peak shape, however, demonstrate that HDC has sufficient sensitivity to allow eventual determination of molecular weight distributions for polyacrylamides. After sonication for 20 sec, the peak maximum in Figure 8(*d*) shifts toward the shoulder. After 50 and 100 sec of sonication, the R_f continues to decrease, indicating further polymer degradation. In addition, the peaks become narrower and more sharply defined, indicating a decrease in polydispersity. At 100 sec of sonication, the relative viscosity of the 200-ppm polymer solution is too low for accurate measurements; therefore, an intrinsic viscosity could not

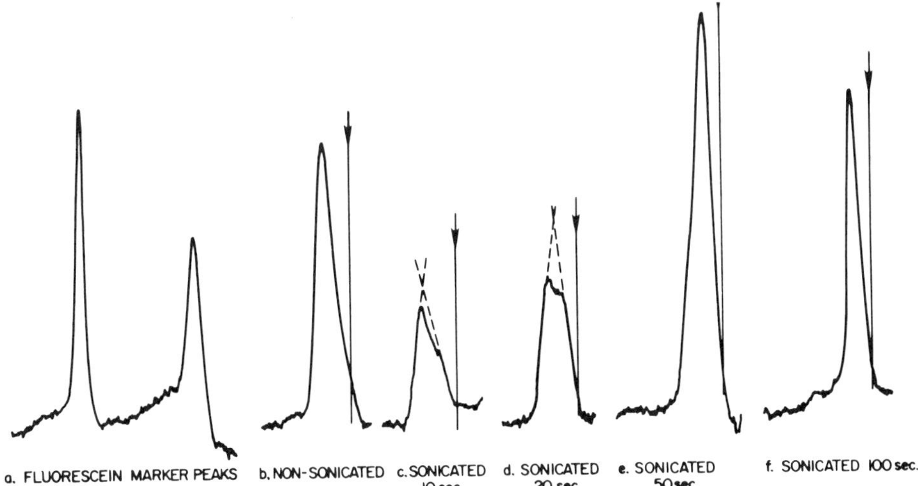

Figure 8. Progression of SF208 degradation. These chromatograms portray the degradation of SF208 by sonication for increasing lengths of time. (*a*) Chromatogram of fluorescein injected in both the sample and marker loops. In *b*) and (*f*) the fluorescein marker peak has been omitted; however, the arrows indicate the location of zero separation. The dotted lines in (*c*) and (*d*) illustrate how the R_f values were measured for these asymmetric peaks.

be determined. The sonication results are in agreement with recent theoretical predictions of polymer degradation which, for central chain scission, show the growth of a low molecular weight shoulder at the expense of the high molecular weight tail, followed by a narrowing of the distribution (32).

4.2.4. *Adsorption on Packing Surface*

An increase in R_f is noted after approximately 20 injections of polyacrylamide samples into the column packed with glass spheres. The extent of this increase can be seen by comparing the R_f of SF208 from Table 2 to the R_f of the nonsonicated sample of SF208 in Table 3. The time span between the two experiments was 3 wk. Increases in R_f were recorded for lower molecular weight samples and latex standards as well. These observations are consistent with irreversible adsorption of polymer onto the glass surfaces of a newly packed column. The adsorption reduces the size of the flow passages in the bed and, therefore, produces an increase in R_f. The amount of polymer that can be adsorbed on the glass surfaces appears limited; after an initial increase in R_f, no further changes are

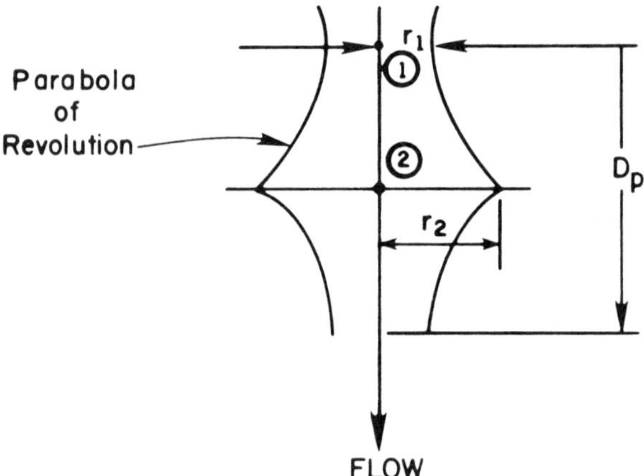

Figure 9. Unit cell of Payatakes' model for porous media (33). Velocities v_1 and v_2 in Equation 4 refer to positions 1 and 2 in the figure.

observed. The results of the degradation experiment show that the ability of the column to separate molecules according to size is not impaired by adsorption, once the active sites on the packing are saturated.

4.2.5. Flow Rate Studies

To determine the extent to which a polymer molecule deforms as it flows through the column, the R_f of the polyacrylamides was measured as the flow rate was varied from 0.38 to 2.35 ml/min. The elongational strain rate at each flow rate can be estimated from the following equation (30):

$$\dot{\epsilon} = \frac{v_1 - v_2}{(D_p/2)} \tag{4}$$

where $\dot{\epsilon}$ is the elongational strain rate and D_p is the diameter of the packing. The velocities v_1 and v_2 correspond to the maximum and minimum centerline velocities in a constricted unit cell model of porous media (see Figure 9), developed by Payatakes et al. (33, 34). The calculated strain rates and measured R_f values are summarized in Table 4. At high strain rates, the separation (i.e., R_f) is independent of flow rate; only at the lowest flow rate, 0.38 ml/min, is a significant increase in R_f observed.

A possible explanation for the lack of sensitivity to flow may be found in a series of articles by Durst et al. (35–37), describing a marked increase in pressure drop during the flow of dilute solutions (50–200 ppm) of par-

Table 4. Flow Rate Studies of SF208

Flow Rate (ml/min)	R_f	$\dot{\varepsilon}\ (s^{-1})^a$	De^b
0.38	1.060	270	5.1
1.20	1.048	830	15.6
1.44	1.052	980	18.4
1.69	1.050	1160	21.8
2.35	1.050	1280	24.1

^{a} Elongational strain rate.
^{b} Deborah number.

tially hydrolyzed polyacrylamides through porous media. This phenomenon was attributed to transition of the polymer conformation from a spherical coil to an elongated ellipsoid, as a result of flow through the converging and diverging passages in the bed. Using kinetic theory arguments, they relate this transition to the Deborah number, a dimensionless stretch rate:

$$De = \dot{\varepsilon} \cdot \tau \tag{5}$$

Here $\dot{\varepsilon}$ is the elongational strain rate and τ is a time constant for polymer deformation that can be calculated from experimentally determined variables:

$$\tau = \frac{[\eta]\eta_s M}{N_{Av}kT} \tag{6}$$

where τ = experimental time constant, $[\eta]$ = intrinsic viscosity, η_s = solvent viscosity, M = polymer molecular weight, N_{Av} = Avodagro's number, k = Boltzmann's constant, and T = absolute temperature.

In their experiments, Durst and his coworkers found that the pressure drop increase occurred at a critical Deborah number of 0.5. Above this value, the excess pressure drop remained constant. These effects were noted for polyacrylamides ranging in molecular weight from 3 to 18 million. It was concluded that the transition from a spherical to elongated conformation occurred when the flow conditions corresponded to a Deborah number of 0.5.

We calculated the Deborah numbers for our column at each of the flow rates studied, and the results are reported in Table 4. The Deborah numbers are well above the critical value of 0.5 for all flow rates. The lowest flow rate, which produced the largest increase in R_f, has a Deborah number that is still an order of magnitude larger than the critical Deborah number. The kinetic theory argument proposed by Durst may explain why there is no significant change in R_f for the four highest flow rates. The polymer molecules may be in a highly stretched state in which their conformations are insensitive to flow rate. Although this explanation is speculative, similar stretch-state transitions have been predicted by de Gennes (38) and observed with birefringence and modeled by Fuller and Leal (39). We are conducting experiments at very low flow rates in an attempt to observe an increase in apparent hydrodynamic volume at De $\approx$ 0.5, as predicted by Durst.

4.2.6. *Ionic Strength Dependence*

It is well-known that polyelectrolyte conformation is strongly dependent on solvent ionic strength (40). The intrinsic viscosity of partially hydrolyzed polyacrylamides decreases with increasing solvent ionic strength, indicating a contraction of the polymer chain (41, 42). To determine if these conformational effects can be seen using HDC, samples of SF208 were examined in mobile phases of varying ionic strength. A sample is prepared by diluting 1 ml of the stock polyacrylamide solution with 30 ml of the mobile phase. The ionic strength of the mobile phase is varied by adding sodium sulfate to a solution of 0.2% sodium lauryl sulfate and 0.002 M sodium azide. One run also includes the mobile phase used in the xanthan experiments. The ionic strength of the eluants are calculated from

$$I = \frac{1}{2} \sum_i c_i (z_i)^2 \tag{7}$$

where I is the solution ionic strength, c_i is the concentration of the ith species, and z_i is the charge of the ith ion. To switch mobile phases, the column is flushed with approximately three column volumes of the new solution. The results of the experiment, shown in Table 5, indicate that the R_f values of SF208 do not appear dependent on ionic strength. In light of intrinsic viscosity experiments (41, 42), the results of the HDC experiments appear anomalous. Durst's theoretical and experimental analysis, however, suggests that the effects on polymer conformation resulting from the hydrodynamics of flow may be much greater than the conformational effects resulting from the ionic strength of the polymer solution. If this hypothesis is true, no change in R_f would be expected.

Table 5. Ionic Strength Dependence of SF208 on R_f

Ionic Strength (M)	R_f
0.01	1.051
0.01 (Xanthan mobile phase)	1.052
0.16	1.053
0.31	1.051

Also, the poor resolution obtained in HDC may not be sufficient to detect differences. Further investigation of these ionic strength effects at flow conditions with lower elongational stresses will be undertaken.

4.3. Dextran

Dextrans are a class of polysaccharides produced by the microbe *leuconostoc mesenteroides* that find application in medicine as blood-plasma substitutes. The polymer is highly branched, and a wide range of molecular weights, obtained by fractionation of native dextran, is commercially available. Very high molecular weight dextran is frequently used to determine the exclusion limit or void volume of SEC columns. The molecular weight distributions of dextrans can be determined readily using either conventional or high-performance SEC packings.

Dextran was selected to evaluate the utility of HDC for polymers of smaller hydrodynamic size than polyacrylamides or xanthan and to show the range of overlap between SEC separations and HDC separations. Studies with this polymer also provide a convenient test of HDC, since dextran samples can be analyzed independently by SEC. As a nonpolyelectrolyte with spherical symmetry in solution (43), dextran possesses, at the same molecular weight, a lower effective size than the two polymers already discussed. The only sample analyzed by HDC had a molecular weight of 2 million, as shown in Figure 10. For this column, the R_f translates into an effective diameter of 0.09 μm. The sharpness of the dextran peak probably arises from a narrow molecular weight distribution. Native dextran has a very high polydispersity (100–1000), and the fractionated sample probably has a smaller, although significant, polydispersity of 2–10 (44).

5. SUMMARY

The results presented in this chapter reveal the considerable promise of HDC for analysis of ultrahigh molecular weight macromolecules. The

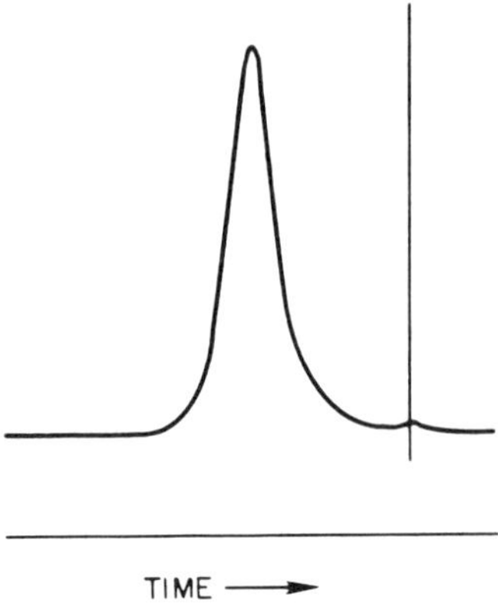

Figure 10. HDC Chromatogram of dextran. For this 2-million molecular weight sample, the size distribution is apparently narrow, leading to a sharp peak similar to the peaks obtained during calibration with monodisperse standards. The conditions for this run are identical to those described in Figure 5, except that a detector range setting of 0.50 was employed.

technique, which can be applied to a variety of water-soluble polymers, is both rapid and inexpensive. With an upper size limit of approximately 1–2 μm, the range of HDC encompasses even the largest water-soluble polymer. The lower size limit is below 0.03 μm. To date, HDC separations of polymers have been used to determine average hydrodynamic sizes, but not molecular size distributions. Because HDC has low resolving power, the determination of size distributions requires deconvolution of the experimental chromatogram to correct for axial dispersion. The high precision of the HDC separation makes successful deconvolution possible, as demonstrated by McGowan and Langhorst (45) for the separations of latex spheres by HDC. They quantitatively analyzed mixed populations of latex spheres using a laboratory computer for on-line data acquisition and analysis. Their techniques are presently being adapted for the analysis of polymer separations in our laboratory. In addition to field flow fractionation (Chapter 7) and ultracentrifugation, HDC is a promising technique for the determination of molecular weight distribution of ultrahigh molecular weight polymers.

Despite the encouragement afforded by the early results discussed in this chapter, significant problems in the field remain unsolved. Polymer degradation induced by high strain rates in the bed, which are of the order $1000 \ s^{-1}$, has not yet been studied systematically. Preliminary studies on polyacrylamide samples showed minimal degradation, whereas substantial degradation of DNA samples was observed (46). Degradation can be reduced by decreasing flow rates, with a corresponding increase in the time needed to perform an experiment. A second problem arises from polymer-packing interaction and its influence on column calibration. Drift in column calibration with time is observed in some cases, which may be caused by adsorbed polymer molecules that alter either the dimensions of the interstitial regions in the column or the interactions between the packing surface and the molecule being separated. Frequently, we found it necessary to perform calibrations before and after each experiment. Even at low polymer concentrations (less than 200 ppm), this effect was noted. To minimize adsorption, surface modified packings or changes in mobile phase composition must be investigated. Finally, simple models of the separation fail to adequately predict the effect of flow rate on the separation. This failure may arise from the high strain rates so far employed—the polymer may be stretched to the extent that other forces are negligible. To increase our knowledge regarding the mechanism of separation, we are considering more realistic kinetic theory models for the flow of deformable polymers in the column. The analysis is complicated by the complex flow fields existing in the bed.

To tackle these problems and expand the current range of HDC, experiments with iron oxide particles, tobacco mosaic virus, and polymers such as DNA are under way. Each of these species has unique conformational properties which may provide insight into the nature of the HDC separation. Furthermore, we are examining chromatographic parameters in an attempt to substantially increase resolution. One can anticipate further applications of HDC as the separation of these model compounds is understood.

ACKNOWLEDGMENTS

We gratefully acknowledge the National Science Foundation (CPE-8003320) for financial support, Dow Chemical Company for a fellowship, and the Exxon Foundation for an equipment grant. In addition, we thank Hercules, Incorporated for fabrication of the particle elutriator and Dow Chemical Company for packing HDC columns. The technical assistance of Drs. Hamish Small and Marty Langhorst of Dow Chemical Company on HDC separations, George Holzworth of Exxon Central Research on xanthan properties, and Howard Barth of Hercules, Incorporated on size exclusion chromatography of water soluble polymers was invaluable. Much of the sample preparation and synthesis were painstakingly performed by Faith Morrison and Richard Williams. Finally, we thank Carol Field for her unflagging patience in typing the many revisions of this manuscript.

REFERENCES

1. S. M. Shawki and A. E. Hamielec, *J. Appl. Polym. Sci.,* **23,** 3323 (1979).
2. F. Lambert, M. Milas, and M. Rinaudo, *Polym. Bull. (Berlin),* **7,** 185 (1982).

3. H. Small, *J. Colloid Interface Sci.,* **48,** 147 (1947).

4. D. C. Prieve and P. M. Hoysan, *J. Colloid Interface Sci.,* **64,** 201 (1978).

5. C. A. Silebi and A. J. McHugh, *AIChE J.,* **24,** 204 (1978).

6. G. McGowan and M. Langhorst, *J. Colloid Interface Sci.,* **89,** 94 (1982).

7. P. J. Flory, *Principles of Polymer Chemistry,* Cornell University Press, Ithaca-London, 1978, p. 606.

8. G. Holzwarth and E. B. Prestridge, *Science,* **197,** 757 (1977).

9. A. Jeanes, J. E. Pittsley, and F. R. Senti, *J. Appl. Polym. Sci.,* **5,** 519 (1961).

10. M. Rinaudo and M. Milas, *Biopolymers,* **17,** 2663 (1978).

11. G. Chaveteau and N. Kohler, SPE 9295, 55th Annual SPE Fall Meeting, Dallas, 1980.

12. G. Holzworth, *Prepr., Div. Pet. Chem., Am. Chem. Soc.,* **21**(2), 281 (1976).

13. S. L. Wellington, *Prepr., Div. Polym. Chem., Am. Chem. Soc.,* **22**(2), 63 (1981).

14. S. L. Wellington, "Enzyme-Filtration Clarification of Xanthan Gum," U.S. Patent No. 4,119,492, October 1978.

15. G. Holzwarth, *Carbohydr. Res.,* **66,** 173 (1978).

16. A. N. de Belder and K. Granath, *Carbohydr. Res.,* **30,** 375 (1973).

17. G. G. Guilbault, *Practical Fluorescence: Theory, Methods, and Techniques,* Marcel Dekker, New York, 1973, p. 292.

18. S. Bakalayar, J. Yuen, and D. Henry, *Spectra-Physics Chrom. Tech. Bull.,* Spectra-Physics Corp., 1976.

19. H. P. Keller, F. Erni, H. R. Linder, and R. W. Fregi, *Anal. Chem.,* **49,** 1958 (1977).

20. I. S. Krull, M. H. Wolf, and R. B. Ashworth, *Am. Lab.,* **10**(5), 45 (1978).

21. R. K. Prud'homme, G. Froiman, D. A. Hoagland, *Carbohydr. Res.,* **106,** 225 (1982).

22. J. G. Southwick, A. M. Jamieson, and J. Blackwell, *Prepr., Div. Polym. Chem., Am. Chem. Soc.,* **22**(2), 143 (1981).

23. P. J. Whitcomb and C. W. Macosko, *J. Rheol.,* **22**(5), 493 (1978).

24. G. Paradossi and D. A. Brant, *Macromolecules,* **15,** 874 (1982).

25. H. Yamakawa and M. Fujii, *Macromolecules,* **6,** 407 (1973).

26. J. B. Hays, M. E. Magar, and B. H. Zimm, *Biopolymers,* **8,** 531 (1969).

27. M. Milas and M. Rinaudo, *Carbohydr. Res.,* **76,** 189 (1979).

28. R. A. Gelman and H. G. Barth, *J. Appl. Polym. Sci.,* **26,** 2099 (1981).

29. F. R. Dintzis, G. E. Babcock, and R. Tobin, *Carbohydr. Res.,* **13,** 257 (1970).

30. R. K. Prud'homme and D. A. Hoagland, *Sep. Sci.,* **18,** 121 (1983).

31. W. W. Graessley, *Polymer,* **21,** 258 (1980).

32. M. Ballauff and B. A. Wolf, *Macromolecules,* **14,** 654 (1981).

33. A. C. Payatakes and M. A. Neira, *AIChE J.,* **23,** 922 (1977).

34. M. A. Neira and A: C. Payatakes, *AIChE J.*, **24,** 43 (1978).

35. F. Durst and R. Haas, *Rheol. Acta,* **20,** 179 (1981).

36. F. Durst, R. Haas, and B. U. Kaczmar, *J. Appl. Polym. Sci.,* **26,** 3125 (1981).

37. R. Haas, F. Durst, and W. Interthal, in A. Yerruijt and F. B. J. Barends (Eds.), *Flow and Transport in Porous Media: Proceedings of Euromech 143,* A. A. Balkema, Rotterdam, 1981, pp. 157–168.

38. P. G. de Gennes, *J. Chem. Phys.,* **60,** 5030 (1974).

39. G. G. Fuller and L. G. Leal, *Rheol. Acta,* **19,** 580 (1980).

40. A. Eisenberg and M. King, *Ion-Containing Polymers, Polymer Physics Volume II,* Academic Press, New York, 1977.

41. T. Schwartz and J. Francois, *Makromol. Chem.,* **182,** 2757 (1981).

42. G. Muller, J. P. Laine, and J. C. Genyo, *J. Polym. Sci.: Polym. Chem. Ed.,* **17,** 659 (1979).

43. K. Gekko, in D. A. Brant (Ed.), *Solution Properties of Polysaccharides, ACS Symposium Series 150,* ACS, Washington, 1960, p. 415.

44. Pharmacia Fine Chemicals, "Dextran Fractions, Dextran Sulphate, DEAE-Dextran; Defined Polymers for Biological Research," 1975, p. 7.

45. G. McGowan and M. Langhorst, *J. Colloid Interface Sci.,* **89,** 95 (1982).

46. K. A. Larson, MS Thesis, Princeton University (1983).

MANUFACTURERS OF INSTRUMENTS FOR PARTICLE SIZE ANALYSIS

Beckman Instruments, Inc., Spinco Division, P.O. Box 10200, 1117 California Avenue, Palo Alto, California 94304.

Brookhaven Instruments Corporation, 200 Thirteenth Avenue, Ronkonkoma, New York 11779.

Cambridge Instruments Ltd., Rustat Road, Cambridge CBI 3QH, England (Cambridge Imanco, 40 Robert Pitt Drive, Monsey, New York 10952).

Chromatix Inc., 560 Oakmead Parkway, Sunnyvale, California 94086.

CILAS—Compagnie Industrielle des Lasers, Route de Nozay 91460, Marcoussie, France (U.S. Representative: Marco Scientific Inc., 1031 H East Duane Avenue, Sunnyvale, California 94086).

Climet Instruments Company, 1320 W. Colton Avenue, P.O. Box 151, Redlands, California 92373.

Cooper Laboratories Incorporated, 3145 Porter Drive, Palo Alto, California 94304.

Coulter Electronics, Inc., 590 West 20th Street, Hialeah, Florida 33010.

Fleming Instruments Ltd., Caxton Way, Stevenage, Hertfordshire, United Kingdom.

Fritsch GmbH, Hauptstrasse 542, D6580 Idar-Oberstein 1 Federal Republic of Germany.

Gaulin Corporation, 44 Garden Street, Everett, Massachusetts 02149.

HIAC/Royco Instruments Division, 141 Jefferson Drive, P.O. Box 2168, Menlo Park, California 94025.

Horiba Instruments Incorporated, 1021 Duryea Avenue, Irvine, California 92714.

Joyce–Loebl Division, Vickers Instruments Incorporated, 300 Commercial St., Malden, Massachusetts 02148.

Ladal Scientific Equipment Ltd., Warlings, Warley Edge, Warley, Halifax, Yorks, United Kingdom.

Langley–Ford Instruments, 29 Cottage Street, Amherst, Massachusetts 01002.

Lars A. B. Ljungberg & Co., Stockholm, Sweden (Manufacturers of Celloscope).

Leeds and Northrup Company, Microtrac Division, 3000 Old Roosevelt Blvd., St. Petersburg, Florida 33543.

Malvern Instruments Ltd., Spring Lane, Malvern Worcestershire, WR14 1AL United Kingdom. (U.S. Represenative: The Munhall Co., 5850 North High St., Worthington, Ohio 43085.)

Micrometrics Instrument Corporation, 5680 Goshen Springs Road, Norcross, Georgia 30093.

Microscal Ltd., 79 Southern Row, London, W10 8AS, United Kingdom.

Nicomp Instruments, P.O. Box 6463, Santa Barbara, California 93111.

Particle Data, Inc., Box 265, 111 Hahn Street, Elmhurst, Illinois 60126.

Pen Kem, Inc., 341 Adams Street, Bedford Hills, New York 10507.

Seishin Enterprise Co. Ltd., Sotobori Sky Building, 13 Honmura-cho, Ichigaya, Shinjuku-ku, Tokyo, Japan.

Shimadzu Scientific Instruments, Inc., 9147H Red Branch Road, Columbia, Maryland 21045.

Simon Carves Ltd., Stockport, Lancashire, United Kingdom.

Sira Ltd. South Hill, Chislehurst, Kent, BR7 5EH England, U.K.

Spectrex Corporation, 3594 Haven Avenue, Redwood City, California 94063.

Technord Associates, Inc., P.O. Box 574, Deerfield, Illinois 60015.

Thermo Systems, Inc. (TSI), Box 43394, 500 Cardigan Road, St. Paul, Minnesota 55164.

TOA Electric Co., Kobe, Japan.

Union Giken Co., Ltd., Osaka, Japan.

Vickers Instruments Ltd., Haxby Road, York, Yorkshire, Y03750, United Kingdom. (300 Commercial St, Malden Massachusetts 02148.)

Wyatt Technology Company (Science Spectrum), P.O. Box 3003, Santa Barbara, California 93105.